"A powerful and essential riposte to the reactionary nostalgia that blights the public discussion of food."

—**George Monbiot**, author of *Regenesis: Feeding the World Without Devouring the Planet*

"Dutkiewicz and Rosenberg have a knack for writing against food orthodoxies in a way that is surprising, refreshing, and informative. Polemical but reasoned and based in the idea that everyone deserves access to food pleasures, this exciting book informs about the food systems we have inherited and imagines the food worlds that could be. A rethinking of common ideas about food that is sure to radically upend how we think about what we do eat, what we should eat, and what we could eat, if only we could redistribute access to the everyday hedonism that we all deserve."

—**Kyla Tompkins**, James Beard award–winning author of *Racial Indigestion*

"Are you tired of half-baked platitudes about the problems with our food system and hungry for real solutions? Then this book is for you. Dutkiewicz and Rosenberg refuse simplistic formulations and offer a complex, incisive, and, yes, provocative analysis, coupled with a rousing call to remake powerful industries to serve humane and sustainable ends. This book is sustenance for your mind as it imagines more democratic and delightful ways we can all fill our stomachs."

—**Astra Taylor**, author of *Democracy May Not Exist, But We'll Miss It When It's Gone*

"*Feed the People!* takes an entirely fresh approach to our 'broken' food system. The two authors are academic experts unafraid to point out the elitist flaws in a number of proposed food system remedies. Their coast-to-coast tour of our nation's foodscape is egalitarian and humane.

They earn our trust by providing useful information in a conversational tone, rather than partisan lectures. Open-minded readers will be rewarded and learn."

—**Robert Paarlberg**, author of *Food Politics:*
What Everyone Needs to Know

"*Feed the People!* cuts through both foodie nostalgia and corporate greenwash to reveal the real politics of what's on our plates. Dutkiewicz and Rosenberg show that the industrial food system is both a triumph of abundance and a site of profound exploitation—and that the answer isn't retreating into elitist fantasies of artisanal purity or storybook narratives of a past that never existed. Instead, they demonstrate how we can demand justice, sustainability, and pleasure—for everyone. This is a bold, necessary book that refuses despair and gives us a vision worth fighting for."

—**Will Potter**, author of *Green Is the New Red: An*
Insider's Account of a Social Movement Under Siege

"*Feed the People!* overturns foodie clichés while taking the politics of food seriously. Dutkiewicz and Rosenberg show how industrial food has made abundance possible and how its strengths can be harnessed for justice, sustainability, and pleasure. With wit, clarity, and urgency, they argue for 'democratic hedonism' as a compelling alternative to both industrial excess and locavore elitism, advocating for policies that enhance worker rights, expand food access, and reduce environmental harm while preserving the pleasures that make eating joyful. *Feed the People!* is essential reading for anyone who cares about what we eat and why it matters."

—**Bruce Friedrich**, founder and president,
the Good Food Institute

FEED *the* PEOPLE!

FEED *the* PEOPLE!

Why INDUSTRIAL FOOD IS GOOD AND HOW TO MAKE IT EVEN BETTER

JAN DUTKIEWICZ | GABRIEL N. ROSENBERG

BASIC BOOKS

New York

Basic Books
Hachette Book Group
1290 Avenue of the Americas, New York, NY 10104
www.basicbooks.com

Printed in Canada

First Edition: February 2026

Published by Basic Books, an imprint of Hachette Book Group, Inc. The Basic Books name and logo is a registered trademark of the Hachette Book Group.

The Hachette Speakers Bureau provides a wide range of authors for speaking events. To find out more, go to hachettespeakersbureau.com or email HachetteSpeakers@hbgusa.com.

Basic Books may be purchased in bulk for business, educational, or promotional use. For more information, please contact your local bookseller or the Hachette Book Group Special Markets Department at special.markets@hbgusa.com.

The publisher is not responsible for websites (or their content) that are not owned by the publisher.

Print book interior design by Sheryl Kober.

Library of Congress Cataloging-in-Publication Data has been applied for.

ISBNs: 9781541603783 (hardcover), 9781541603790 (ebook)

MRQ-T

10   9   8   7   6   5   4   3   2   1

Contents

THE FOOD SYSTEM PARADOX

HAVE YOU EATEN AT YOUR LOCAL WAFFLE HOUSE LATELY? ANTHONY Bourdain once remarked that Waffle Houses make some of the best damn waffles around. The waffles at Waffle House are light and crispy, the sugars in the batter caramelized by the griddle to a deep, sweet amber. When drizzled in syrup and topped with a fluffy dollop of butter, a single bite delivers a glorious trifecta of sweet, salty, and creamy. Wash it down with hot coffee in that signature mug. It isn't the best coffee in the world, but it isn't the worst, either. And the refills are usually free. Since Waffle House opened in 1955, it has served close to one billion of those waffles.[1]

You might agree with us right away. Yes, you love those waffles. Or maybe they're a guilty pleasure, a memory you've tried to convince yourself that your taste buds have outgrown. Or maybe you're turning your nose up at the very thought of stepping foot in a Waffle House. But why?

Don't dismiss these pleasures just because they're simple and common. Although you might be able to reproduce those waffles on your own with a bit of practice and effort—or maybe you're the breakfast hero at home, a batter-and-griddle maestro who can outshine them—the reliable pleasures Waffle House delivers are a wonder owed to the scale and precision of the modern food system. And here's the thing. The waffle you make in your own kitchen? Or that elaborate one with lavender buttercream and black-currant syrup at the trendy brunch spot? Both of those will almost certainly piggyback on the same crops and modern industrial technology as the Waffle House, but the difference is that although you can make a few waffles (and a huge mess) in your own kitchen, and the brunch place can make a few hundred, Waffle Houses turn out on average 145 waffles every minute of every day all year long, and they do it with an inexpensive and dependable consistency that delights the droves of ordinary people who pack their booths 24 hours a day, 365 days a year.[2] That accessibility ensures that you're likely to see a diverse crowd of people dining there: goth teens hitting vapes and dodging parents, inked-up hipsters heading home from the club, truck drivers taking a break from a long haul, hungry workers between shifts, and families in their Sunday best eating after church. Waffle House is simple, it is magnificent, and its food has been perfected through hundreds of millions of orders.

Waffle Houses are the product of a food system that, for all its warts—of which there are many—manages to democratize and scale access to food and its many pleasures in a way unparalleled in human history. It is a food system of incredible abundance. Of shelves reliably stocked with varied products, food that is safe to eat, and cheap and accessible nutrition for many people. Even if your go-to isn't a late-night waffle but an early-morning green smoothie, the technologies and systems that make it possible are for the most part the same. In fact, you are more likely to find the raw ingredients for a better future for the food system at the Waffle House than you are at your local farmers' market.

If the idea that the food system is, for the most part, working strikes you as controversial—or even heretical—you are not alone. An entire food-writing industry exists to convince you otherwise. Much that is written about food today presumes that Waffle Houses, and modern food-production technology and food policy and eating habits more broadly, are the reason, not the solution, for our food ills. Most food writers suggest that the modern food system needs to be discarded and replaced with one where consumers search for locally sourced, labor-intensive, and cooked-from-scratch options, supporting the small, alternative, and artisanal purveyors who represent a pastoral ideal of sustainable food. Inconvenient? Too expensive? Too tired from work to cook for yourself? Quit your whining and do your chores.

But consider that a single Waffle House waffle would be a remarkable achievement in most human societies throughout the ages. The Mesopotamians, Aztecs, Romans, and countless others, though brimming with brilliance, creativity, and ingenuity, did not produce a single waffle. Billions of waffles? An unimaginable achievement in any era other than our own. In premodern Europe, milling flour by hand was time-consuming and difficult labor, usually done by women. Larger mills, powered by water, wind, or animals, were controlled by the wealthy. The finely milled flours needed to produce a delicate and fluffy pastry were expensive and rare, and most people ate coarse and rough loaves when they ate bread. Industrial milling, first powered by steam and then by electricity, made finely milled flours much more widely available, even as improved agricultural productivity dramatically reduced the price of the wheat from which the flour was milled. With the addition of commercial leaveners like baking soda in the twentieth century, the fluffy waffle that had once been the pleasure of only princes is now ubiquitous. And the higher universal standard of living of our time, related in large part to most people not having to work in agriculture, makes those waffles widely affordable and accessible.

Yes, Waffle House has serious problems. It doesn't pay its workers well enough, its menus are crammed with unhealthy options, and it serves up food that is produced in ways that damage the environment. For some, Waffle House might represent something like the epitome of brokenness: a low-paying, high-fructose, lowbrow, all-monocrop disaster. For these haters, there is no place in the food system for the Waffle House.

But poor worker pay, unhealthy options, and outsize environmental impacts are problems that afflict not just Waffle Houses but most restaurants and supermarkets in the United States, including, often enough, the expensive and artisanal ones. It is precisely these problems that need to be addressed, one by one, to improve our food system rather than abandoning the whole model in favor of nice-sounding but ultimately dubious alternatives.

But doing that means digging into the details of how the food system actually works. Into the environmental impact of the production of different products and what sustainable alternatives exist. Into the relationship between minimum wages and food insecurity for workers, and how they organize for better pay and working conditions. Into some basics of nutrition and food processing, which will tell you that having a few of those waffles now and then is both good for the soul and not all that bad for you. And into which consumer actions and which government policies can help us keep the best parts of the current food system while jettisoning the bad ones.

Our vision for a better food system entails keeping what works and doing away with what doesn't. So, despite their problems, Waffle Houses inspire us because they offer a vision for food grounded in popular, accessible, and abundant pleasures for all. What people love about Waffle House is also what they love about most of the items served up by the food system as it exists today, in a restaurant or at home, whether it's burgers and fries, mac and cheese, or a kale salad. We are trained to think of fresh and healthy ingredients as alternatives to what the conventional food system offers, but at scale and at an affordable price point, they are utterly dependent on the same technologies and techniques that make waffles so cheap. The products of the conventional food system are often snobbishly dismissed as the frivolous and empty delights of junk food, but this book shows that nearly all food we eat is the product of the modern "industrial" food system, junk food and healthy food alike. We want to take the best parts of that system and harness them for a broader political transformation.

We come not to bulldoze the Waffle House but to liberate it. The current system directs much of its extraordinary productivity and efficiency into a surplus of profits and delectable treats for a select few players. Let's change the equation. This book lays out a blueprint for how that same productivity can serve the environment, human and animal health, food access, and labor justice, providing accessible and sustainable pleasures for everyone.

THE MODERN FOOD SYSTEM PRESENTS A BEWILDERING PARADOX: It's never been better at feeding humanity, and it's never exacted a more catastrophic toll. For millennia, most people eked out bare subsistence through grueling agricultural toil. Hunger, malnutrition, and famines were frequent. By contrast, food today is cheap and plentiful. Not only does this allow people to live longer, happier, and healthier lives, but, liberated from the plow, billions of people can enjoy the comforts and amenities of modern life. A world of backyard barbecues, art museums, medical breakthroughs, and, yes, waffles wouldn't be imaginable without the incredible productivity, abundance, and convenience unlocked by the industrial food system.

Yet the way we eat is also a massive problem. Many components of the modern food system are optimized to put short-term profits above anything else, overworking people, animals, landscapes, and our climate—producing enough food to feed billions while also lining the pockets of corporations and governments. Food and agriculture alone account for nearly one-third of all global greenhouse-gas emissions. The demand for crops to feed animals and humans drives deforestation: The world loses ten million hectares of forest every year, an area the size of Kentucky. This drives species to extinction and reduces our planet's capacity to capture carbon. Food production also sucks aquifers dry while chemical runoff poisons waterways.[3]

The effects on humans aren't much better. Although modern agriculture produces more than enough protein and calories to feed everyone, these are unevenly distributed. More than 40 million Americans are food insecure. Seventy percent are overweight, 40 percent are obese, and 10 percent have diabetes, driving skyrocketing health-care costs. Meanwhile, the same agricultural expansion that is driving global climate change also supercharges the risk of zoonotic illnesses, with concentrated animal feeding operations (CAFOs) acting as toxic breeding grounds for swine flu, avian flu, and antibiotic-resistant bacteria. The people toiling in agriculture are among the most underpaid and mistreated in the economy, and they have some of the highest rates of workplace injury and abuse. The food system is at once the most abundant it's ever been and, at the same time, is *the* crisis of our time, an interlocking set of problems that touches nearly everyone.[4]

You may already be familiar with the many problems with how food is produced and consumed. The media are filled with stories of the food system's failings and the near-ubiquitous statement that "the food system is broken." Scores of bestselling books, including Michael Pollan's *Omnivore's Dilemma*, Alice Waters's *We Are What We Eat*, and Wendell Berry's classic *The Unsettling of America*, have all advanced similar big-picture indictments of how American food is produced and how Americans eat.[5] For the most part, they've also converged on a familiar solution: Scrap what we've got and start over. Go small, local, and organic to eat your way to a better food system. This approach, sometimes called farm-to-fork, slow food, or locavorism, suggests that consumers can spark broader transformation of the food system by focusing their purchases (and diets) on real food produced the right way, the way it used to be before industrialization and corporatization. They should give up processed junk and eat only whole and fresh ingredients produced through traditional methods on small farms in and around their community and lovingly prepared by artisan chefs and home cooks.

But this approach gets both the past and the future wrong.

A halcyon age of sustainable farming and clean eating never existed. If your great-grandparents were like many people in the United States a century ago, they may well have struggled with food insecurity and serious nutritional deficits. In fact, one of the greatest challenges faced by human societies throughout history has been providing enough food to feed everyone, and when they did avoid famine, even preindustrial societies often used agricultural methods that overtaxed workers and the environment. Although the current food system has many problems, it's also an improvement for most consumers and eaters over what came before. More importantly, solutions to the problems that do exist—and yes, there are many—must start with a fair assessment of both the current food system's faults and its benefits. Real, scalable, feasible solutions are right in front of us. We just need to understand what they are and how to make them work.

THE FIRST STEP IN FIXING THE FOOD SYSTEM IS GETTING AWAY FROM SIM-plistic claims that it is broken. Much food writing and media coverage of food and agriculture is based on this idea, and given the litany of issues we've outlined with how we produce and consume food, this might seem like an obvious conclusion. But while we agree that there is much wrong with the food system, "broken" is not the correct word to describe it. And it's not just the word. It's a particular way of thinking that leads to that claim. It's like telling a handyperson your house is broken when the air conditioner is on the fritz and you're sweating through the couch. It doesn't convey what components are malfunctioning and for whom this is a problem. In this simple example, fixing the HVAC makes your house comfy and livable again. Saying "my house is broken" won't get the HVAC up and running, and the claim itself instills a sense of listless dread instead of offering a solution to fix what's not working. So too does saying that the food system is broken suggest impending peril; it offers no real vision of a better future and only vague gestures at systemic change. We suggest an alternative. To analyze the benefits and failings of the food system—and ask what's broken and what isn't—requires grappling with this complexity and engaging in what academics and policymakers aptly call food systems analysis.

To do that, we need to do what most of those who claim that the food system is broken don't often do, which is explain what exactly the food system is. Not to be too pedantic about it, but the food system is, first and foremost, a *system*. It's a network of interconnected but disparate actors and institutions working in concert, but without centralized coordination, to achieve the aim of making food available to consumers.

For the Waffle House to serve up waffles, there is no central Department of Waffle Administration or even all that many people interested in waffles, but rather an entire constellation of different actors who play a part in making sure that all the elements are in place for the waffle to end up on your plate for the few minutes—seconds even—it lasts before you scarf it down. Some of those people—like the truck driver hauling flour down the interstate or the bureaucrat implementing zoning laws for restaurants—may not even know they are in the business of making waffles. It might be tempting to see the food system as a simple process of farmers producing food and retailers and restaurants buying it and then reselling it to eaters, but this would be a gross oversimplification. In fact, the food system is what scholars refer to as a complex system, namely one with relatively simple individual components (seeds, plants, farmers, food companies, commodity brokers, civil servants) but extremely complex overall behavior. And it is a system that is dynamically interwoven with the other systems that make up our complex society. That's why, for example, federal-government decisions about energy policy determine why so many Midwestern landscapes are dominated by vast cornfields grown for ethanol.

And in a capitalist world, a motivating and exacerbating factor for the benefits and ills of the food system is the fact that almost every single actor throughout the food value chain is trying to make a profit. This motivation is often concealed with high-minded rhetoric about feeding the world or looking out for communities or the environment, but any clear-eyed analysis of the food system as it exists must be serious about economic motivations.

There are many ways in which food systems experts have visualized the food system and its constituent parts. Some imagine it as a chain (or a cycle) moving from production through processing, distribution, and consumption (farm to table). Others, like the United Nations Food and Agriculture Organization, divide up the food system into two blocks, the food supply chain and the consumer-food environment (behind and after the farm gate), and examine how these are affected by environmental and socioeconomic factors. Other models abound, and many have merit. But the one we prefer comes from Nourish, a food education initiative.[6] For our purposes, we've simplified the Nourish model a tad to arrive at four subsystems of the food system: biological,

9

economic, political, and social. For a taste of how such analysis works and what it can tell us, let's get back to that Waffle House waffle.

The main ingredient in Waffle House waffles is wheat. Although wheat is one of the few crops grown in every state in the United States, there are certain climate and environmental conditions that it prefers. This is where the biological system comes in. It includes things like the quality and health of soil, the amount of rain and sunlight available to grow crops over a year, the biodiversity of farming and nonfarming species of plants and animals, and the presence of pollutants (from farming, industry, or urbanization) in a given area. These determine what can be grown, how it can benefit from local conditions, and how it might adversely affect the local environment.

Wheat grows best in loamy soil made up of sand, silt, and clay, and although it requires a steady amount of water to grow from seed to full maturity over about three or four months, it also prefers long hours of sunlight and moderate temperatures between 65° and 75° Fahrenheit. In the United States, this has made Kansas and North Dakota the ideal locations for vast wheat fields, which cover, respectively, 15 percent and 17 percent of these states' entire territory. And growing vast amounts of it involves monocropping (growing only one crop at a time in large quantities rather than mixing crops), including the use of artificial fertilizers and pesticides that harm bird and mosquito species, and digging wells or diverting water through vast irrigation systems. Increasingly, the limits of such efforts are evident. Much farming on the semiarid Great Plains depends upon water drawn from the Ogallala Aquifer, a vast underground lake that stretches from Nebraska to Texas. All this irrigation means that farmers have been drawing down the aquifer faster than it can be replenished, and many ecologists warn that it will be unable to support large-scale agriculture in the coming decades. As reservoirs like the Ogallala dry up and as soils degrade, the biological subsystem of the food system is now one of its most fragile components.[7]

Paying attention to how biological factors influence the food system underscores a key point: There is little that is "natural" about any food system, and especially not ours. Any food system entails some effort to shape, redirect, contain, and sometimes destroy the biological systems that human societies confront. Nor is "naturalness" something that always makes for a better food system. Foodborne illnesses are perfectly natural, and efforts to limit them involve technology and human intervention, but we humbly suggest that a food system with less botulism is superior. But it's not just the environmental aspects of food production that are not "natural." In the market for wheat and waffles, there is little that is natural about supply and demand.

The economic system is the value chain that produces and delivers your food. This is the network of private actors trying to turn a profit or earn a wage from farm to fork. This starts with farmers and farmworkers (an important distinction we'll get into in a later chapter). But it also includes grocery stores, trucking companies, restaurants, and large food corporations, such as those that produce inputs like fertilizer and machinery like tractors, and also Wall Street traders who buy and sell everything from food commodities and shares in food companies to agricultural land.

In the case of the wheat that winds up in your waffle, the seeds a farmer uses might come from Bayer, the German pharmaceutical company that bought agricultural biotechnology company Monsanto in 2018. Bayer could also sell the farmer the pesticide to be sprayed on the crops. Once it grows, the wheat might be harvested with a John Deere combine. The wheat will then likely be sold to a grain-elevator operator such as Archer-Daniels-Midland Company (ADM) or perhaps a large cooperative like Agtegra. From there the grain will be sold and processed into flour by a flour mill such as C.H. Guenther, which makes all the wheat flour for Waffle House waffles. And those waffles will be cooked on specialty waffle irons made by Wells Manufacturing of

Smithville, Tennessee. Meanwhile, the price that farmers are paid or that millers pay elevators depends on a host of factors, including the price of wheat futures on the Chicago Mercantile Exchange, which is often determined as much by how good the weather and climate have been during a growing season (there's that biological subsystem) as by the analyses and sentiments of investors in global markets, such as when Russia's invasion of Ukraine sent wheat prices on global markets rising precipitously on fears of shortages.[8]

The point here isn't just that food value chains are complex and that a wheat seed from Fairmount, North Dakota, will pass through many hands before it becomes a waffle on a plate at 2 a.m. in Decatur, Georgia. Rather, it's that relationships of power exist within the economic subsystem that have important effects on the food on your plate. Economic actors can, for instance, exert pressure on the political system. By lobbying politicians at all levels of government, including through organized producer checkoff programs like the Kansas Wheat Commission and the Texas Wheat Producers Board or trade groups like the National Restaurant Association (of which Waffle House is a member, and which fiercely opposes increases in the minimum wage) and through direct donations (Waffle House's biggest political donation in 2022 was to the National Republican Senatorial Committee), economic actors can push for preferential policies and lobby against laws and regulations they see as threatening to their business model.[9]

But the biggest impact of the economic subsystem is actually getting food in front of consumers, into their shopping baskets, and onto their dinner tables. With farmers deciding what they can produce profitably, companies developing new products or new takes on old ones, retailers and restaurants shaping what people can buy with their decisions about stocking shelves and crafting menus, and food companies flooding media and public spaces with advertising (the fast-food industry spends more than $5 billion per year on advertising in the

United States alone), the economic actors that make up the food system don't just feed us; they also directly shape what we eat and, in many instances, what we want to eat.[10]

The profit motive is so obvious you can (and they did) fairly title a documentary *Food Inc.*, but it's not a complete story because, in countless and often unseen ways, big political decisions shape both corporations and the food they put on our plates. The political system entails all the regulations, laws, taxes, and subsidies that govern food production and consumption. They span the gamut from crop insurance subsidies to the Supplemental Nutrition Assistance Program (SNAP, formerly known as the food stamp program) to expert panels that develop national dietary guidelines. The political system contains all the institutions that push, pull, and mold food production, incentivizing some forms of farming and consumption and disincentivizing others, and it ultimately determines what food ends up on our dinner tables and if that food is safe to eat.

Laws and regulations touch virtually every aspect of Waffle House operations. From the municipal zoning rules for where restaurants can be located to Food and Drug Administration (FDA) requirements in the Food Code designed to minimize the risk of customers getting sick from faulty food storage or preparation, virtually every aspect of our food system is shaped by policy. (This is also the reason that most restaurant kitchens around the country, no matter the cuisine or location or price point, look so similar to each other.) But some regulations, and especially those concerning labor—the line cooks who make those millions of pancakes and the serving staff who deal with hungry, hangry, and outright offensive customers—vary widely from state to state. In Georgia, where Waffle House is based, the state minimum wage doesn't apply to tipped employees, meaning that a Waffle House line cook might make $7.25 per hour while a waiter or waitress starts at $2.13 and must earn their living through tips, off the goodwill of customers.

But policy and marketing can do only so much. Corporate belly flops like New Coke and the failure of government attempts to get people to eat more fruit and veggies both show that much about what we want to eat is determined by personal taste, context, heritage, and tradition, all of which are shaped by the social system in which we live.

The social system is probably the primary way you interact with food. It shapes what you find delicious, disgusting, and desirable in a meal. The social system is, quite simply, the information, norms, and beliefs about food that are all around you and inform what you want to eat. It comprises the social and information networks, both physical and digital, where we find knowledge, identity, and culture, and that we use to guide our daily choices. It draws from our personal, familial, and national histories, including the stories we tell about what foods are deeply meaningful for us and which ones are just junk or fads. It includes formal education acquired through schools, but also informal information gleaned from TikTok, much of which is hard to distinguish from the many half-truths and lies with which we're bombarded incessantly through food advertising. The social system also includes many ideas about food that we simply take for granted and that predispose us to believe that our particular foodways are superior. Crucially, this involves ideas about food ethics; about what constitutes good, natural, or "real" food; and about the appropriate ways to change the food system or keep it the same.

Waffle Houses are one variation of a cherished staple of American food culture that sprang into existence during the twentieth century and that is tied to our nation's similarly distinct love of automobiles: the roadside diner. Usually located at the intersection of major thoroughfares, roadside diners offered food to the growing number of automobile and truck drivers zooming along the dense network of highways crosscutting the US countryside and, notably, built by massive public works campaigns before and after World War II. Because these drivers were usually strangers to the places where they stopped, and in a great hurry to boot, diner menus tended to prioritize simple, affordable, and homey fare that could be cooked (and eaten) quickly and without fuss or pretension. Breakfast staples, such as scrambled eggs, hash browns, bacon, pancakes, and, most importantly, waffles, were a perfect fit, a mass food culture that would appeal equally to someone from North Dakota or North Carolina. It's simply impossible to imagine Waffle Houses having the same cultural salience in a country where the love of driving vast distances for long stretches was less common than it is in the United States.

Poke someone enough about why they eat what they eat, why they crave what they crave, and chances are that they'll admit that it is, in the end, somewhat arbitrary. The contingencies of cultural context and personal history play an enormous role, much more so than rational decision-making. This means, in turn, that appealing to rational decision-making to generate better outcomes can take us only so far; people's deepest desires and assumptions are daunting challenges to even the most carefully crafted and rational schemes to change what they eat. You may not like the Waffle House. You may think the world would be a better place without it. But here's the dish: No matter how you feel about it, the earnest and powerful love that other people have for the Waffle House is an empirical reality with which you must deal, and one no less powerful than the composition of soil, the price of wheat, or laws setting the minimum wage.

The Waffle House waffle contains multitudes. Each one is a small miracle that requires countless ingredients from across the food system to come together. This is all complicated! And coupled with the scale of agriculture and its myriad impacts, it's a lot to take in. The food system is an infinitely complex and interesting subject, but it's also easy to get overwhelmed. Revealing this complexity is the first part of using food system analysis to understand how our food system works. But if all the food systems approach did was show just how complicated everything was, it would not be helpful or an improvement over what you find in most food writing. However, food systems analysis allows three important things.

The first is to show that claiming that the food system is "broken" is nonsensical. To address the benefits and harms of particular aspects of the food system, you need to be specific. The second is to recognize that there are levers to be pulled in many places that can change what we put on our plates. The food system is not one single thing that you choose to take or leave, but rather a series of things open to intervention, change, and improvement. The third is that in working toward a better food future, you can't start from scratch; you have to start in the here and now. This doesn't mean that all possible food systems lead to Nestlé, Monsanto, and Waffle House or that ours is the best version of the food system we can hope for. Certainly, all waffles all the time would dull our taste buds and thicken our waistlines. We're all for options. But thinking more holistically about what exactly goes into a food system reveals that escapist foodie fantasies are a dead end: neither feasible nor scalable nor desirable as the basis for reliably feeding all of society.

And, much like we can't start from scratch or go back in time to some imagined better age when food and agriculture were great, we can't pin our hopes for food system change on any one solution. Perhaps even more overused than the claim that the food system is broken are claims that if we just did one particular thing, be it regenerative cattle ranching or eating local or embracing some new technology, we could save the food system or even the world. But the food system is not one broken thing that needs fixing; it is a complex puzzle with myriad options for improvement and rearrangement.

That locavore dream, the small organic farm, fails to address countless aspects of the food system, including treatment of workers and food access for the hungry and food insecure in a country of 330 million people on a planet of more than 8 billion. These are complex questions that can't be dismissed with a handwave or that will simply disappear if we graze cows on sunny pastures and buy heirloom radishes. We'll confess that we love fresh produce prepared with care and the odd trip to the farmers' market. But foodie fantasies have created a smoke screen over viable solutions to the food system's problem.

The same goes for the Waffle House. We can't just wave a magic wand at it. For the Waffle House to achieve its delicious promise, it'll need many fixes, big and small. Increasing the minimum wage would do wonders for many of Waffle House's workers. Introducing meat alternatives to the menu would make it more sustainable and healthier. Growing wheat that is more resilient to drought using the latest gene-editing technology would protect waffles from price and supply shock in a future altered by climate change. The list goes on. Each problem must be addressed separately, with the understanding that there might be no single ideal solution.

The standard we use for judging solutions in this book is quite simple: They have to actually address the problem, they must have a provably large potential impact on the problem, they must be feasible to be implemented at scale, and their benefits in one part of the food system have to outweigh their potential negative impacts elsewhere. And here's the kicker: We have all the right ingredients, and the recipe is hiding in plain sight!

FOOD SYSTEMS ANALYSIS GIVES US A POWERFUL TOOL FOR UNDER-standing what's for dinner, and it can show us where the food system has gone awry, but it cannot tell us, by itself, how we should change the menu. It gives us cooking techniques, but it doesn't give us a recipe to follow. And recipes for fixing the food system cut to the heart of politics: interests, tastes, values, and beliefs.

And that's where we turn to a guiding concept of this book and of our vision for a better food system: democratic hedonism. We'll talk at greater length about this idea in the next chapter, but for now we'll offer this definition: Democratic hedonism is an approach to politics that sees moral value in the simple pleasures that people experience in their daily lives, and it favors political, collective, and institutional actions to expand access to those pleasures. The goal of this approach is to figure out how to reduce harms and maximize pleasures for as many people as possible. Pleasure, and its promise, is one of the things that defines a good life, and it can motivate and sustain political action, arming people with the fortitude that change requires. When it comes to food, pleasures may be personal and sensual, but they are also frequently social and communal, and, regardless, even solitary pleasures, as food systems analysis shows, are reliant on other people (as well as on plants and animals).

Hedonism alone we find unattractive. Have you ever had the misfortune to sit next to a rich wino at a bar or fancy restaurant? This is a very wealthy person who, not content to wallow in average alcoholism, has transformed their drinking problem into an expensive (and irritating) hobby, drinking only the rarest and most expensive wines, expounding on tasting notes, and paying closer attention to the vintages on offer than to their dining companions. Now consider the Waffle House, where the pleasure is accessible and you're not going to wax lyrical about the terroir of the syrup. You might, instead, fill up on calories and caffeine for your road trip, laugh with friends, or plot a book about the food system. Let's shift from the rich wino's selfish hedonism to the democratic hedonism of the Waffle House. Democracy is an ethos and system for sharing power and social goods broadly and equitably. From this perspective, pleasure needn't be a selfish or isolating obsession, a portal to narcissism or self-indulgence. Rather, making sure that everyone has abundant access to pleasure should be a social project that inspires collaboration, care, mutuality, and solidarity.

Put that way, the delicious promise of democratic hedonism aligns with a key concept that scholars use to study food systems: food security. Food security is at the heart of most progressive food politics, from local urban food programs to the United Nations' global Sustainable Development Goals. A community is food secure if it has regular access to enough safe and healthy food to meet its nutritional standards. In academic terms, food security requires food availability, access, and utilization. Combine these three ingredients with environmental sustainability and labor justice to create our recipe for a better food system. Think of it as a democratically hedonic layer cake.

The first layer of a better food system is availability. Is there enough food produced to meet the nutritional needs of a country, community, family, or individual? In the United States, availability isn't so much the problem. There's enough food in the country to feed everyone a few times over. The problem is that what is produced and by whom doesn't always strengthen food security; to the extent that the food system provides food security, it is often in spite of glaring and obvious problems. Understanding availability means looking at who produces what, where, and why, which also clarifies how and why we might change it.

The second layer is sustainability. Agriculture is one the biggest users of land and fresh water on the planet, as well as being a major emitter of greenhouse gasses. What we produce—how we get food availability—should ideally be as environmentally benign as possible, but that also means changing what food we demand and eat. From farms to food technology to efforts at dietary change, keeping agriculture within planetary limits and minimizing its environmental harms while providing food security ensures not just a more livable planet but also the ability to produce food and provide food security well into the future.

The third layer is access. Can people get the food that's available? At the most basic level, this depends on whether people can afford the food they need. And if they can't do it on their own, are there institutions that provide them with food, be it government programs that supplement their incomes or charitable organizations? These questions intersect with more complex ones that are not strictly about food, such as wages, rent prices, and even the physical landscape that people must navigate to obtain food.

The fourth layer is labor. Who are the people bringing food to our plates? How are they treated, and how much are they paid? Most importantly, can they afford to eat, and eat well? This spans the laws, policies, labor disputes, and worker organizing that determine the conditions in which the tens of millions of workers in the food system labor. Those workers are also eaters, which means that attending to their food security requires better pay and working conditions.

The fifth layer is utilization. Once people have availability and access to food, are they eating a nutritious and delectable diet? This is an even broader question. It depends on the quality, safety, and nutritional value of the food itself. But it also depends on whether people are buying nutritious food and cooking it properly. If you're already at the supermarket and can afford food, but you're walking right past the produce aisle and bakery to get a Coke and chips, we're not going to judge you, but that's a utilization issue. Utilization also raises many of the most rancorous debates in food circles, including about issues like processed food and fast food and about heavy-handed government intervention, such as regulating food supplements and taxing sugary drinks.

The recipe we propose for a better food system is divided into chapters addressing each of the layers above and can be boiled down to this: Food security, done sustainably, while improving the lot of workers will make for a better food system for more people.

Between the two of us, we have been studying the American food system for more than three decades. Gabriel is a historian who teaches at Duke University and writes about the history of agriculture, policy, and science and how all have shaped American culture and landscapes. He grew up in Indiana, and before he was a professor, he worked food services jobs, including as a cook. Now he splits his time between Durham, North Carolina, and Berlin, Germany, where he is a Senior Research Scholar at the Max Planck Institute for the History of Science. Jan is a political scientist whose research focuses on the environmental, ethical, and public-health dimensions of food production and food policy. He grew up between Warsaw, Poland, and a small village just outside it amid socialist (and then postsocialist) hobby farms before moving first to Canada and then to the United States. Jan is a contributing editor at *The New Republic*, where he often writes about the politics of food, and is a contributing writer at *Vox*, for which he has covered food-related issues including the pro-vegan activism of the animal rights group PETA, the science of comparing different foods' environmental impacts, and the farmer-led protests that shook Europe in 2024. Together, we have written numerous essays about food for *The New Republic*, *Vox*, *The Guardian*, *Dissent*, and other outlets. Jan is a vegan. Gabriel is not.

Over the six chapters that follow, we will take you on a tour of American food, but maybe not the sort you're used to. Chapter 1 explains what most American food writing gets wrong, including its weird and off-putting take on food pleasure. It then offers a taste of democratic hedonism as a palate cleanser. This is a chapter of critique and theory offered as a conceptual appetizer. But if you're eager to bite right into the meat of the matter, you can move right on to the next chapters. Each of these cuts into the layers of that democratically hedonic layer cake.

Chapter 2 takes us to farms in Iowa and the nitty-gritty details of the business of agriculture to explain how, why, and by whom most crops are grown in the United States. In Chapter 3 we tackle the uncomfortable topic of how to reduce the environmental impacts of food production, focusing on meat and dairy, foods that are both abundant and harmful pleasures. We find promising low-tech and high-tech solutions being developed in California, the nation's largest agricultural state. Chapter 4 moves east to New York City and Durham, North Carolina, to look at public programs, from SNAP (food stamps) to school lunches to food banks, which ensure that all people, no matter how poor they may be, have access to a good diet. Chapter 5 follows the thread of economic justice to the South and back to Waffle House, where workers are fighting for better wages and to improve the restaurants where they work. In Chapter 6 we head to Boston to talk about food utilization with nutrition and public health experts, examining some foundational myths and misconceptions about diets, nutrition, and food processing. These misconceptions have important implications for both your personal diet and for how policies and politics can facilitate good health outcomes while respecting personal choices.

At a basic level, the aim of this book is to help you better understand the food system and to arm you with insights about how to navigate it and how it might be improved. Some of the information in this book will be useful to you as you make personal decisions about what to eat; some of it won't. But even when we can't help you be a more discerning eater, we think this book will empower you as a citizen. What we choose to eat can be part of an effective politics of food, especially when it is reinforced by collective action, electoral politics, well-crafted laws and regulations, and smart institutional design. We hope that the book teaches you about the opportunities for and challenges of addressing the food system's many problems and inspires you to get involved in working to fix them.

We want to offer you a different way of thinking about food and its role in our society, one that is responsive to a now-global political crisis in which the basics of democratic pluralism are increasingly called into question. If Americans cannot share meals together, it is hard to imagine that we can live and work together, much less share a national government.

We hope this book inspires your curiosity, not just about where your own food (and food pleasures) come from but also about the way others interact with and enjoy food. Curiosity, we believe, is a prerequisite to the collaboration and solidarity that will be necessary for building a world of abundant, accessible, nutritious food pleasures for all.

Chapter 1

THE CASE FOR DEMOCRATIC HEDONISM

A GOOD FOOD SYSTEM SHOULD BE BASED ON FOOD THAT TASTES good, is available to as many people as possible, and causes as little harm as possible.

This simple fact was made abundantly clear to us one chilly March morning in Brooklyn, when we feasted on breakfast sandwiches from ATM, Jan's favored deli near where he teaches at Pratt. The night before, we had scored a reservation at Eleven Madison Park, a legendary fine-dining restaurant in Manhattan awarded three Michelin stars that had gone plant-based in 2021, leading to much international debate, publicity, and acclaim. The multicourse meal had its high points to be sure—a crisp and salty sweet potato tart and the between-course mugs of rich umami roasted rice broth stuck with us—and the restaurant's veggies-only menu tugged at our political and environmental affections. We wanted to love our experience. What could be better for two professors who write about the many harms the food system does, including to animals, than dining at perhaps the world's most celebrated vegan restaurant? The menu was thoughtful, meticulously curated, Instagram-ready, duly expensive, and, in its own way, environmentally and politically righteous.

What it wasn't was particularly pleasurable. The dishes were cerebral odes to technique and restraint that, sadly, just didn't deliver. One memorably catastrophic dish featured a piece of steamed Chinese broccoli entwined in a lonely, wan noodle. These listlessly rested in a puddle of a murky soy sauce concoction that someone had dusted with shaved truffles. The one promising item on the plate—the noodle—was so miserly in its portioning that it felt downright hostile. The whole dish was completely overwhelmed by the truffles, which tasted more like an idea of luxury than luxury itself. Most of the meal wasn't that bad, but the dish did illustrate the limits of foodie aesthetics and ideals in stark terms: theoretically noble and high-minded ideas that work best on paper above practically sensual delights that work on the plate, asceticism masquerading as minimalism, and appeal rooted in expense and elitism. Rarified pleasures that, for the most part, aren't all that pleasurable. The service, we should note, was indeed world-class.

But the next morning, Jan's stomach roiling from one too many too-fancy cocktails, Gabriel's eyes still bleary from sleep, a single bite of ATM's plant-based breakfast sandwich, a greasy delight modeled on the standard-issue New York City deli bacon, egg, and cheese sandwich, and the meal from the night before was a distant memory. This is what we'd wanted all along. Simple, savory, familiar, filling, affordable, and, in *its* own right, environmentally and politically righteous.

Food is one of the most reliable and quotidian sources of pleasure, and as the variety of available food has multiplied and its cost has plummeted over the past half century—inflationary run-ups here and there notwithstanding—it is also now surely one of the most affordable and common pleasures. And this is a great thing. One of the best things about the modern world, we would argue.

What would a politics rooted in delivering abundant, accessible food pleasures entail? And how would it differ from the vision offered by most foodie writers and thinkers?

The foodies get neither the empirics of agriculture nor the concept of pleasure right. To make that case, this chapter closely engages with one of the most significant food writers in the history of American letters and the primary theorist of foodie pleasure. The Kentucky farmer, poet, and essayist Wendell Berry inspired a revolution in food writing and is a touchstone reference in nearly every major foodie text.

If you've never read (or heard of) Wendell Berry, you have almost certainly read a book, seen a movie, engaged with a social media post, or eaten a plate of food that was directly or indirectly shaped by his writing. Berry's writings helped to launch both the farm-to-fork culinary movement and New Food Writing, the renaissance of journalistic books published since the 1990s that have sought to diagnose the ailments of the contemporary food system and usually dismiss industrial, "fast" food as worthless garbage corrosive to both America's teeth and its soul. That's why if you flip through the pages of books by the journalist Michael Pollan, the food writer and restaurant critic Mark Bittman, the celebrity chef Alice Waters of Berkeley's Chez Panisse restaurant, the activist and founder of the slow-food movement Carlo Petrini, or nearly any other book written about better food and a better food system, you'll find numerous references—most of them glowing, some of them overwrought—to Berry. Waters writes that Berry's writing "electrified" her, with lines that "reverberated deeply and articulated for me a fundamental truth about how I want to live my life." Meanwhile, Pollan credits Berry with forging "a new, more neighborly conversation between American environmentalists and American farmers, not to mention between urban eaters and rural food producers."[1]

High praise from the heavyweights of American gastronomic letters. Too bad Berry got it wrong.

WHAT FOODS GIVE *YOU* PLEASURE?

Consider some common food delights (or feel free to make your own list of favorites): a slab of meatloaf glazed in Heinz 57; a Taco Bell bean burrito (no onions); a double scoop of pistachio ice cream; a sliced ripe tomato with a sprinkle of salt and a dash of olive oil; a strawberry Pop-Tart; a salad of crisp lettuce topped with seared tuna, cucumber, and sliced avocado; fried chicken and a biscuit; silken tofu swimming in Lao Gan Ma chili crisp; a plump peach, juicy and sweet; a handful of tart cherries; a Waffle House waffle dripping with syrup.

What unites these is that each of them is usually (or in some cases always) a pleasure furnished by the modern, conventional, industrial food system. Yes, you can find an expensive and local artisanal version here and there, and, heck, you may even have eaten a sliced tomato you plucked from your very own vine. Ah, but do you also have an olive tree? Did you harvest, mash, and press those olives yourself before you drizzled the oil? No?

That's because, unless you live somewhere where all those items are grown *and* are always in season, you access food pleasures the same way most other people do: Ingredients are brought to you from all around the country and the world, and you purchase them at a grocery store or restaurant. When you or a restaurant combines these, nearly every dish is a mishmash of ingredients, few of them local, many of them packaged, canned, preserved, processed, or premade. A favorite comfort food, a simple pasta marinara, uses boxed pasta, canned and peeled tomatoes, pressed and bottled olive oil, cheese (or a plant-based equivalent), and maybe a few leaves of local, fresh basil. You may justly rhapsodize about the earthy garden-grown herb, but the satisfaction of that local basil depends on the other ingredients: the sturdy, reliable, and affordable mass-produced boxed pasta and canned tomatoes furnished by the industrial food system. Less glamorous? Perhaps. But without them you'd just be eating a handful of basil leaves. The strength of the industrial food system lies in precisely the many basic pleasures that you take for granted so much that you barely notice that they make up the majority of the dish.

By contrast, haute cuisine is an arena for the fetishization of food, chefs, and, to some extent, eaters. Fine dining is rooted in the premise that products and palates are best when they are rare, rarefied, and refined. Don't believe us? Try arguing with your friends about which of the dishes at Barcelona's Disfrutar, the number one restaurant in the world in 2024, is the most transcendent; don't blame us if your friends throw a Miller High Life at you. That's because having those kinds of opinions about cuisine requires training, commitment, and investment that must be nurtured by masterful chefs conjuring exquisite flavors for the select few. An entire industry of restaurants, culinary schools, and food media exists to reinforce the separation between dining that is fine and dining that is common, with the Michelin star system its seal.

Never mind that the sorts of aspirational tastes modeled by the fine-food world often rest on a faulty foundation. For instance, American wine judges have been shown in numerous blind taste tests to be consistent in describing the same wines in the same way only about 10 percent of the time.[2] In a 2001 French experiment, many members of a panel of wine experts couldn't even differentiate between a white and a red when the white was colored with red food dye.[3] That these tastemakers manage to keep churning out wine rankings and musing on tasting notes is a testimony to the power of marketing and the industry's need to maintain an illusion of enlightenment more than any actual superiority of gustatory sense.

Haute cuisine is a realm of fetishization in another sense as well, focusing only on the food to the exclusion of anything else. It ignores everything from animal suffering to labor abuses. Unsurprisingly, scandals about the mistreatment of workers by megachefs and the faking of ingredients proliferate, such as the infamous case of the Willows Inn on Lummi Island in Washington state, where not only was the local food not local, but wage theft and other abuses were also so rampant that the $500-per-meal restaurant ended up paying out $600,000 to former employees after a class action lawsuit.[4]

But as far as it has no pretensions to being a model for how everyone should eat, fine dining, when it doesn't venture into fraud or labor-code violations, is annoying but mostly fine, a pleasure left to those to whom it appeals and who can afford it. Eleven Madison Park is, for many, a beloved pleasure, and a far more righteous one than Willows Inn or any number of pretentious steakhouses whose only claim to fame is "elevating" their cuisine and their prices by performatively sprinkling everything with coarse salt and whose best dishes are nowhere near as good as Eleven Madison Park's. (In August of 2025, Eleven Madison Park announced it was ending its plant-based experiment.)

The problem is that much earnest thought about remaking the food system and the American diet tends to, often despite itself, aspire to haute cuisine. One could easily miss the appeal of common food pleasures when reading most contemporary food writing. Crack open one of dozens of foodie books about making a better food system, and chances are you'll be met with good intentions wrapped in elitist austerity, elitist judgment, or both. Such writing assumes that only food grown or prepared the "right way," with time-consuming, handcrafted care and freshly harvested artisanal ingredients from small local farmers, *tastes better*. Here, transparency and ostensibly good politics end up, despite themselves, dovetailing with the snobbery of haute cuisine.

This is the fundamental conceit at many of the most world's most celebrated farm-to-table restaurants, such as Blue Hill at Stone Farms, located on a small farm up the Hudson from New York City, as well as Copenhagen's legendary Noma, the restaurant of chef Rene Redzepi that has long been considered among the world's finest. Barber, Blue Hill's chef, takes hyperlocalism to one extreme by growing most of the food on the farm where the restaurant is located. Redzepi goes to another extreme by prioritizing ingredients foraged from the Scandinavian landscape that he meticulously transforms through byzantine preservation and fermentation techniques. It's hard to say if the epicurean delights at these restaurants live up to the hype: Both, armored in sterling reputations for their enlightened food sensibilities, are exclusive and forbiddingly expensive.

That's horseshoe theory for foodies: Local, artisanal, heirloom, and craft have become the cuisine of the global jet set. Farm-to-fork, slow, and locavore chow is often sold with a self-righteous rhetoric of egalitarian access, communitarian heart, environmental awareness, and a love of humble, simple ingredients prepared with unpretentious care and grown by real farmers. Yet the restaurants that have perfected this cuisine are so pretentious, elaborate, and aestheticized that precious few people can ever afford (or would even want) to eat there with any regularity. These chefs have cultivated a local garden, but it is guarded by a high fence and private security. And the lucky few who can afford those meals may fly thousands of miles from places like Austin and Menlo Park so that they can eat a sweet potato grown on-site or a pile of seaweed someone dredged out of the bay. We don't put much stock in the idea of "food miles," but it would be interesting to add up the carbon costs of the customers' "people miles" at these elite hyperlocal destinations.

Even if we allowed that the politics of this style of cooking were better—sometimes, but only sometimes, it is—this story about flavor and pleasure is still suspect. Many contemporary food writers will sneer that the supposed pleasures many people take from a greasy breakfast sandwich or the salty crunch of a Dorito are just cheap and empty thrills, worthless and trivial. Would that pleasure only came from what was righteous, like a tomato bought from the farmers' market, the path to heaven would be a Sunday stroll. If you've ever tried the organic health food Doritos knockoffs sold at Whole Foods, you know it just ain't so; sorry, but Doritos are delicious.

We're not saying you should gorge yourself on junk food or, conversely, stop going to your favorite restaurant or artisan bakery. But deriding the conventional food system and the food pleasures it provides, out of an elitist and aesthetic scorn or, worse, misguided political righteousness, misses much that is good and must be protected about the food system. Even if one wants to change the food system. In fact, especially if one wants to change the food system. And if one wants to understand why the vaunted food revolution of farm-to-fork has failed to change the food system, it's in part because it tried to dismiss and replace the real and abundant and affordable pleasures that people take from industrial food rather than trying to make that food better, healthier, more sustainable, and less exploitative.

Common food pleasures fortify our lives and human relationships. Feeding others and eating with them is an important way in which we become aware that our personal sensual delights are shared and possible because we live together, as social creatures. A coffee with a friend, a slice of pizza and a beer after work, a dinner party meal cooked in anticipation of great conversation and laughter, or even a lunch made from thrown-together leftovers so you can make it to your kid's soccer game. Even dining alone on a favorite meal can be joyous. But taking common food pleasures seriously requires that we can consider how to expand access to them, how to deepen their meaning, and how to invest ourselves in the pleasures of others even as they become invested in ours.

These common—in both senses—pleasures are behind the idea

that drives us in writing this book: democratic hedonism. We borrow this concept from the Yale professor and political theorist Joseph Fischel, who writes, in the context of sex, that "democratic hedonism is not a facile celebration of more people getting off more of the time" but, instead, a call to "think more boldly and strategically about democratized access to pleasure and intimacy."[5] We think that the concept applies just as much to food, so, with Fischel's blessing, we use it and develop it here on our own terms: as a political project aimed at building a yummier society, one that can accommodate food pleasures as diverse as our society.

We stand by pleasure as a political and policy orientation—that's what makes us hedonists—but we believe individual pleasure should be seasoned with an egalitarian ethos that mutually invests us all in each other's pleasure. Simply put, hedonism needs democracy to avoid the pitfalls of selfishness, narcissism, and snobbery. Democratic hedonism affirms pleasure as one of the things that fills life with joy and makes it worth living. We cherish pleasure and believe a politics without it will be unattractive, drab, listless, and doomed to failure. But the point of democratic hedonism is not to ignore the harms that may accompany pleasure; it's to figure out how to reduce the harms and maximize the pleasures for as many people as possible.

ON THE SURFACE, WENDELL BERRY WAS ALSO ARDENTLY COMMITTED to the pleasures of eating. In fact, his most famous and oft-quoted essay is titled just that, "The Pleasures of Eating." But it's really more of a list of chores. The gist of it is that most people are what Berry terms "industrial eaters." They eat mindlessly, without reflection, knowledge, or responsibility, and they want to eat without working very hard to feed themselves. When honest work is replaced by convenience and labor-saving technology, Berry says, you will eat whatever the industrial food system provides even when it provides you with garbage. These industrial eaters are "victims" of their own sloth and ignorance and of corporate agribusiness: "The ideal industrial food consumer would be strapped to a table with a tube running from the food factory directly into his or her stomach."[6]

Berry juxtaposes the emptiness of this industrial eating with what he calls "extensive pleasure," and it's here that we get a real peek at where Berry thinks the worthy pleasure of food resides: not in the experience of eating, but in work! Extensive pleasure is not wedded to the sensual, tactile, or gustatory—what Berry terms the pleasure of the "mere gourmet"—but emerges from "one's accurate consciousness of the lives and the world from which food comes."

Consider Berry's program for how the "industrial eater" can obtain extensive pleasure, which he dubiously asserts "is pretty fully available to the urban consumer who will make the necessary effort." Want pleasure? Get to work:

1. "Participate in food production to the extent that you can."
2. "Prepare your own food."
3. "Learn the origins of the food you buy, and buy the food that is produced closest to your home."
4. "Whenever possible, deal directly with a local farmer, gardener, or orchardist."
5. "Learn, in self-defense, as much as you can of the economy and technology of industrial food production."
6. "Learn what is involved in the best farming and gardening."
7. "Learn as much as you can, by direct observation and experience if possible, of the life histories of the food species."

These may sound good in principle, but what would they mean in practice?

You could go all the way and follow Berry's example. After all, there's nothing more local than homegrown. Berry inherited his farm, but you'll need to purchase a few acres before you can grab a hoe and start tilling. Although we find the valorization of subsistence farming by affluent Americans distasteful, some hardy trailblazers have launched a modern "homesteading" movement largely based on this premise. Most readers will find this option a bit extreme (or flatly impossible), so we'd direct you to the next best thing: the farmers' market.

Be prepared to get up early on the weekend to scour your local farmers' market for the freshest ingredients before the other early birds pick your favorite stall clean. Check the available balance of your bank account before you go; it's going to be much pricier than what you're used to at the supermarket. Supermarkets source their produce from all over the country and the world. That allows them to harness efficiencies of scale and capitalization, the comparative advantages of crops grown in climatically suitable locations, and low prices furnished by a competitive marketplace with tens of thousands of producers.

The produce at a farmers' market, by contrast, is drawn from a local area (usually driving distance to the market) where farmers likely operate (on average) much smaller, less capitalized farms and grow crops that may or may not be suited to the climate. Meanwhile, because there are fewer farmers to choose from, sellers at farmers' markets are shielded from price competition, which means demand can quickly outstrip supply. That leads to markups.

As an aside, local in theory is not always local in practice. Take Belcampo Meat Co. Founded in 2012 by alternative-food entrepreneur and Berry adherent Anya Fernald, it touted itself as a hyperlocal, "pre-industrial brand." With its own farm and artisanal slaughterhouse in Siskiyou County, close to Mount Shasta, offices in Oakland, and stores in San Francisco, Palo Alto, and Los Angeles, the company leaned into high-priced, artisanal meat, urging customers to eat tail to snout, chug bone broth, and feel good about it. Belcampo ticked all the boxes: It boasted about the transparency of its value chain, its treatment of animals was certified humane, and it was an early adopter of the "regenerative agriculture" label. The problem was that this whole Berryesque performance was just that: a performance. On May 26, 2021, an employee at its Santa Monica location posted a video on Instagram allegedly showing the inside of the store's fridge, whose shelves contained conventionally raised (i.e., industrial) beef and chicken not sourced from Belcampo's farm. The whistleblower claimed that the company was buying tenderloin for $10 per pound and reselling it as if it were their own for $47.99 per pound. If there was anything bespoke about Belcampo, it was their bait-and-switch tactics. "Their shit's not local," the whistleblower states matter-of-factly in his video, a tattooed arm holding up a plastic-wrapped tenderloin clearly labeled as a product of Tasmania.[7] In the world of foodies, where aesthetics and reputation are everything, Belcampo couldn't survive, soon shuttering its storefronts.[8]

But presuming you're dealing with honest actors, and you've got the cash, you're still not ready for the farmers' market. First, it's time to hit the books. You'll need to know about crops, livestock, soil, climate, machinery, pesticides and herbicides, fertilizers, and a host of other complex topics. And that's just the big picture. The specifics of how farming works in your area matter even more, as do the identities of the actual farmers. Beyond what crops and how they grow them, you'll need to know how they treat their workers and animals. And you'll need to discover what it means to be an organic farmer, something defined by regulations and law, versus a "regenerative" farmer, a popular term with no agreed-upon common meaning. All these issues will prompt other, more complicated questions that will require serious thought and research to resolve, such as: Just how "local" is a farm that purchases its inputs (seeds, fertilizers, tractors, etc.) in a global marketplace?

Once you've located a suitable local farmer, your research doesn't end in your quest to "prepare your own food." What does it mean, for instance, to avoid eating anything "your great-grandmother wouldn't recognize as food," as Berry adherent Michael Pollan advised? What exactly *did* your great-grandmother eat? Do you have to cook it the way she did too? Did she have a gas range or a wood-fired stove? How do you even cook turnips to be tasty? Is it OK to purchase the rabbit already dressed, or do you have to skin and gut it? Is it still OK to use Himalayan pink salt? What about store-bought pasta? Or do you have to make your own? How long will it take to prepare it all? Can you squeeze it in between work, your kids' dinnertime, and the evening glass of wine (is it local?) paired with a book about fixing the food system? Will your children even eat it? Should you spank them with a wooden spoon like great-granny would have if they don't? Or maybe you should have your kids working in the fields to harvest the wheat for that pasta if you want the full old-timey effect?

The questions add up. Answering them could take some serious sleuthing and careful contemplation. We're skeptical that an explosion of thoughtful choice making about every aspect of eating will make most people happier. A spiraling menu of consumer options when it comes to food, often burdened with serious moral and political significance, sets up more and more opportunities for people to regret what they have chosen. Information overload transforms the "omnivore's dilemma," as Pollan defined the problem of deciding what you should eat, into a labyrinth of stressful, high-stakes questions that rarely have simple, clear-cut answers. For example, no amount of consumer self-education can resolve whether, when the two conflict, you should prioritize farms that use sustainable methods above those that pay their workers fairly. Interpretive problems like that don't resolve to decisive yes-or-no answers but lead to more questions, many of which are valuative, abstract, and irresolvable.

Like Berry, you may be tempted to dismiss as lazy and reckless the people buying Costco rotisserie chickens and ordering Americanized Pad Thai from Uber Eats. But Berry's attitude is snobbish and self-defeating. Even if we agreed that people *should* spend more time conscientiously shopping and laboriously cooking in ways Berry would endorse, there will be no grand food systems transformation if people simply can't and won't do what they ought. Dismissiveness about material barriers like scarce time and money suggest that people like Berry lack a realistic theory of change. Educating consumers about the virtues of homegrown food is lovely, but the hard realities of budgeting tend to teach consumers another lesson altogether. Extolling a virtuous path that no one else has any hope of following is a hallowed American tradition; the early residents of New England preached the predestination of the elect, and locavorism tastes like someone has rebottled that old wine.

Taking all this into account, it shouldn't surprise you that many of the most prominent foodie opinion leaders who champion Berry's farm-to-fork model operate luxury restaurants. Barber's Blue Hill restaurant, for example, will set you back $398 per diner, and that's before wine and the mandatory "22 percent administrative fee." By comparison, Waters's Chez Panisse is a steal at only $175 for a meal.[9] Why are these meals so expensive? In large part, it's because you're paying someone else to do all that legwork we described above, what Berry sees as the true source of food system pleasure. Locavorism done right is extremely time- and labor-intensive for you, or it's expensive because it's time- and labor-intensive for someone else. To his credit, Barber is candid about Blue Hill being a luxury operation and not a model for how most people can or should eat. But that, right there, gives the lie to the whole conceit.

That's why to the extent that the foodie movement has accomplished much of anything, it has mostly shaped the food aesthetics of high-end consumers, the menus of the expensive restaurants that cater to them, the products and advertisements of the upmarket supermarkets and brands that sell to them, and the content of the food pages of newspapers and magazines. This is not inherently bad. We'll confess that we love fresh produce prepared with care. We'd wager that Waters cooks heirloom carrots that are, to quote chef-turned-rapper Action Bronson, "fucking delicious." Done well, the locavore food aesthetic and flavor is as defensible as any other, and there are elements of it we enthusiastically support. We just don't think it's up to the challenge of offering a scalable alternative to the status quo.

Meanwhile, whether people will, in fact, experience more pleasure from completing Berry's list of chores is an empirical matter. As college professors who teach classes about food, we find that students' experiences vary in whether they enjoy learning about where their food comes from *and* whether learning about it alters what they like to eat. Now, we firmly believe people should learn more about the food system or we wouldn't be writing this book. But Berry is confusing two things we prefer to keep distinct: pleasure and accountability. One is phenomenological, a question primarily answerable by those who experience it; the other is a relational question of ethics and politics. Whether you enjoy eating veal (maybe) is a different question from whether you should eat veal (no).

But when Berry is stripped of his high-minded platitudes, his chief anxiety is about ignorance, and his vision of pleasure is rooted in the satisfactions of work and knowledge, not the satiety of food. You might be looking for lunch, but Berry is hunting for meaning. "The thought of the good pasture and of the calf contentedly grazing flavors the steak," he writes. Does it? If you took pleasure from the veal's savory juices and crispy rendered fat or, for that matter, from the salty crunch of a Dorito, you are, according to Berry, suffering from food false consciousness, and your pleasure is degenerate and shameful—a "degraded, poor, and paltry thing." Whether you think that veal or a Dorito is good is a separate issue from whether the sensual experience of eating it gives you pleasure.

What is odd to us is the tendency, on full display in Berry, to try to invalidate and disavow pleasures that motivate eaters' behavior. Berry's approach to all this reminds Gabriel of a certain family member who, when Gabriel came out as gay, told him he just hadn't met the right girl yet—an empirical denial of Gabriel's pleasures and desires masking what was a moralistic judgment of them. Now enjoy the whole quotation from Berry that we gave you only a small taste of before, and the version most food writers don't cite: "*Like industrial sex* [emphasis added], industrial eating has become a degraded, poor, and paltry thing." What exactly is "industrial sex," we wonder? In that line, as elsewhere, "industrial" for Berry and his followers is mostly a moral, not an empirical, category: The key trait of both industrial sex and industrial food is that they are the sex and food that Berry doesn't like.

This makes for great sermons, but it hasn't in fact spawned a mass movement to materially transform the food system, and it's not hard to see why: Eating the locavore way is expensive and time-consuming, and most people don't find it pleasurable enough to justify those costs.

WE STARTED WITH HIS ESSAY "THE PLEASURES OF EATING," BUT WENDELL Berry's most influential work is his book *The Unsettling of America*, published in 1977 by the environmental organization the Sierra Club. It is a powerful and often eloquent polemic against what Berry saw as the decline of the traditional, holistic, and ecologically informed model of farming that had, he believed, predominated in the United States for most of its history. The model described by Berry revolved around a single family carefully tending to and passing on a small plot of land over many generations, father to son to grandson and so on. Invested in the well-being of future generations, farmers would avoid laborsaving shortcuts and unsustainable bonanzas that drew down the wealth of their land, including the fertility of its soil and its biodiversity, such that their posterity could no longer farm it.[10]

This model *settled* America, according to Berry, embodied as it was from the beginning in the Jeffersonian ideal of the yeoman small farmer. That ideal traveled west with the waves of independent small-holders who turned prairie and scrub forest into rich agricultural lands. America's *unsettling*, by contrast, was the uprooting of those small family farms by large-scale commercial operators after World War II.

To put things right, Berry argued that farming should be once again organized almost entirely around small farmers who grew and sold food directly to local communities. He believed that farmers should avoid the seductions of the farming taught at land-grant colleges and pushed by the US Department of Agriculture (USDA). That disastrous path, paved with empty promises of modern technology, extractive efficiency, and short-term profitability, led away from a traditional emphasis on nurture, balance, and farming in nature's image and replaced it with specialized expertise, technology, and inputs that threatened the independence of farmers, bankrupted their soil, and guaranteed their eventual dispossession and immiseration.

Farmers would need to return to "the model nurturer . . . the old-fashioned . . . ideal of a farmer," but they couldn't do it alone. The same processes that had twisted farmers into crude exploiters had also distanced ordinary Americans from where and how their food was grown. This, Berry believed, coarsened the diet, soul, and body alike. This thin relationship to food is what Michael Pollan would later term America's "national eating disorder" in *Omnivore's Dilemma*, a book peppered with paeans to Berry.[11]

The Unsettling of America is an intentionally polemical text—an entertaining and eloquent one at that—and not a work of polished scholarship. It is often beautiful and potent. Its mix of the apocryphal with the factual and the figurative with the literal is the power and license of the poet that Berry is. These excesses would be forgivable if subsequent writers treated the book's generalizations and hyperbole as creative license rather than as documentary realism. But Pollan, for example, gushed in *The Nation* that Berry's analysis in *Unsettling* was so astute that, looking back on his own writing in *Omnivore's Dilemma*, "as a young writer coming to these subjects a couple of decades later, I was rather less original than I had thought."[12]

Stated simply and without the poetry, Berry's ideas sound less appealing and often a bit silly.

Let's begin with the idea that farming was once pursued in nature's image uncorrupted by the desire for filthy lucre. That past is largely a myth.

Not only was the "settling" of America rooted in the land grants that dispossessed Native Americans of their land, but even small American farmers were already integrated into various commodity markets and state- and national-level value chains by the beginning of the nineteenth century, producing surplus to feed growing urban populations. And there is scant evidence that smaller farmers were more sensitive to soil conservation and biodiversity. In fact, small farmers who faced declining soil fertility often lost their farms or simply relocated farther west, taking up plots of land opened by the Homestead Act, the intercontinental railroad, and the federally overseen genocide of the land's prior occupants. The "family farming" practiced there was less intended to protect soil fertility and more to capture, transport, and retain the number one input on those farms—labor—which is precisely why, when social reformers at the end of the nineteenth century turned to the problem of child labor, they discovered that the most intensely exploitative child labor was found on farms and was mandated by the parents of the children. Regardless, the precarious resulting homesteads—around half of all farms created by the Homestead Act failed by 1900—were not famous for their ecological sensibility. Historians such as Donald Worster argue that the ecological and economic catastrophe of the Dust Bowl in the 1930s was caused in part by the awful land-use management strategies of the small-scale, preindustrial family farmers who settled the High Plains.[13]

In fact, histories of antebellum agriculture, like Steven Stoll's *Larding the Lean Earth*, show that the farmers of the period most concerned with conservation tended to be wealthier Northerners and Southern plantation owners. Their favored improved husbandry involved labor-intensive dunging of the soil or the application of imported bat-guano fertilizer, methods that were too expensive for most small farmers.[14]

It's also telling that despite his focus on the merits of traditional farming, in *Unsettling* Berry has little to say about Indigenous modes of farming, Indigenous land management, and Native Americans in general. You would not know from Berry's account, for example, that Indigenous communities across North America practiced complex forms of agriculture that included the use of fire, polycultures, crop rotation, and irrigation, and that they had done so, in many landscapes, for thousands of years prior to the arrival of European colonialism and small commercial farms. Nor would you know that the same Indigenous communities, rather than being rooted exclusively in subsistence farming or hyperlocal economies, had extensive trade networks that crisscrossed the continent and extended into Central and South America.

Rather, Berry's image of American Indians is one of noble but hopelessly doomed dupes who were defeated "not by loss in battle, but by accepting a dependence on traders that made necessities of industrial goods," a description that astonishingly whitewashes the violent role of settlers in the dispossession of Native peoples. In *Unsettling*, that history is perversely used as a pithy illustration of the harms of industrial agriculture on small farmers *now*. Multinational corporations, insatiable urban consumers, and enabling government officials "have made 'redskins' of our descendants, holding them subject to alien values, while their land is plundered of anything that can be shipped home and sold," complains Berry.

We recount this history not to replace Berry's celebratory moralism with our own condemnatory moralism. It is, rather, a plea for contemporary readers and writers to stop uncritically swallowing fantasies of a preindustrial Garden of Eden made up of small, self-reliant, and goodly family farms.

This nostalgia also results in much backward reasoning on the topic of agricultural technology. Berry seems to believe that because his old-fashioned farmers were more righteous, they chose less technologically sophisticated farming methods. Therefore, the use of more advanced technologies by contemporary farmers is a mark of their moral corruption. Berry doesn't seem to consider the possibility that farmers in the past used less sophisticated technology less because of their virtue and more because they did not have access to better technology. As the historians Emily Pawley and Ariel Ron both point out, such farmers were actually often hungry for technologies that substituted capital inputs for labor, readily embracing the "improved" and "scientific" agriculture of the period when they could. In other words, Berry sees unchosen poverty—or at least unchosen labor—and mistakes it for disciplined righteousness.[15]

Contrary to Berry, less *work* need not necessarily mean less *nurture*; less work can make more time for nurture. And lost in Berry's moralism is the fact that the use of technology is not intrinsically moral or immoral: It's how you use technology that matters.

When you dig beneath the rhetorical celebrations of ecological holism, you find that much like Berry smears pleasures he doesn't like as "industrial," so too does he dismiss modes of farming he doesn't like as "industrial." And so too does he dismiss . . . cities.

"Burn down your cities and leave our farms, and your cities will spring up again as if by magic. But destroy our farms and the grass will grow in the streets of every city in this country," famously asserted William Jennings Bryan, a three-time losing presidential candidate, agrarian populist, and all-around dour Christian moralist and (Democratic) party pooper. You'll recognize Berry's view in his words: Cities depending on farmers is good, moral, and natural, but farmers depending on the wares of factories, the currency of governments, or the appetites of urban consumers is unnatural and degenerate.[16]

Berry's advocacy for a transition to more labor-intensive and local forms of farming in the name of the moral purity and independence of farmers is innately anti-urban. Less technology would require vastly more land worked by many more people. It is already impossible for sufficient agricultural land to be found close to major population centers. In a world in which all food was produced on small "local" farms, New York City, Los Angeles, Houston, and a thousand more cities would all either have to be starved into ruins or systematically depopulated, options that would also devastate the industrial and technological bases of our society. Even a less dramatic shift in that direction would make food less available because there was less of it, and less accessible because it would be more expensive. Many people would find this less pleasurable. Notably, the diets of most Americans in the nineteenth century, including the huge portion of people who lived on farms, were less varied, more bland, featured *less* fresh and healthy produce, and generally resulted in worse health outcomes, including malnutrition and foodborne illnesses that are unheard of in the modern era.

In other words, by returning us to premodern agriculture, Berry would return us to shorter, harder lives filled with worse and less food and more and harsher manual labor.

Some fans of Berry, such as the sociologist and farmer Chris Smaje, candidly admit this. Smaje, the author of *A Small Farm Future*, argues that the world is hurtling toward such a severe ecological catastrophe and population crash that a return to local, labor-intensive food economies is unavoidable; the collapse of the food system and urban civilization will produce what policy and polemics cannot, and there's nothing anyone can do about it. Indeed, Smaje *does* advocate for the planned depopulation of large urban centers. We respect Smaje for being candid about his views, but his reasoning, like Berry's, is exactly and obviously backward: Avoiding ecological catastrophe is vital precisely *because* it would return so many people to the poverty of the past. Any solution that slashes agricultural productivity produces the very illness it purports to cure.[17]

It's not surprising that the locavore food revolution never arrived. The consequences of increasing food costs, to say nothing of a shrugging acceptance of the depopulation of most of the world's major cities, is so unpalatable, and the backlash to it would be so swift and severe (think of the American public's furious response to food-price spikes in the past five years), that labor-intensive localism has little purchase in food-policy debates beyond lofty rhetoric and some misguided philanthropic ventures. What it does do is provide the raw ingredients for countless foodie books. But even Berry's most enthusiastic disciples, and those who repeat his mantras for consumers, usually blink when it comes to practically translating his fundamental production principles into large-scale food system solutions. Pollan's *Omnivore's Dilemma*, for example, credits Berry extensively as an inspirational source of a better food ethic, but it has virtually nothing concrete to say about what policies and politics would actually generate a better food system. Similarly, *Animal, Vegetable, Junk*, by cookbook-writer-turned-foodie-author Mark Bittman, exhibits a hostility to the "junk" churned out by the industrial food system, but when Bittman reaches for models for effective food activism, most of his examples come from the efforts of wage laborers on conventional farms and at fast-food restaurants and grocery stores that are intent on engaging and improving the existing industrial food system they work within, not uprooting it. However, in Berry's world—and the one that Bittman ostensibly agrees is a desirable model—none of those jobs would even exist.[18]

But few leading food writers or academic proponents of backward-looking agroecological models of food production grapple with these inconsistencies or the broad food system consequences of their preferred visions. Instead, many just wave away the bad histories and pragmatic infeasibilities. They imagine a world of better food from small farmers and never quite get past the idea itself or, like Pollan with his paragon of better farming, rancher and author Joel Salatin, they hold up fringe characters with quixotic ideals whose production models will never scale. The focus for many contemporary foodie writers is on the food and not the feasibility. That Berry's work is reduced to this idea by his followers is both fitting and ironic, because for all its complaints about the ills of the modern world, including consumerism, it ends up being about diet. It's Weight Watchers for the soul.

BERRY AND HIS MANY FOLLOWERS—BE IT WITH PENS OR CHEF'S knives in hand—have suggested ad nauseam that ordinary people are brainwashed by agribusinesses and fast-food restaurants. They suffer from food false consciousness. The sensual delights they derive from, say, a Waffle House waffle are shallow, empty, and beneath respect. We say that those pleasures are real and compelling, which is why, when they are also harmful, they can be disastrous. In contrast to Berry, we think taking other people's pleasures seriously, even and especially when we don't share them or when they are harmful, is the first step to identifying satisfying but less harmful alternatives.

We stayed with Berry not because he is personally villainous but because his way of thinking has proved so influential among those who influence how we think about food. Even if you've never read Berry, surely you recognize all the ideas we've discussed thus far. His work reinforced an uncritical nostalgia for a past to which we cannot (and should not) return. He fetishized agricultural and culinary naturalism in ways that make practically engaging with useful technologies and production at scale impossible. And his shrill moralism about other people's delights resulted in a way of thinking about gustatory pleasures that is counterproductive.

As far as we're concerned, we ought not abandon the pleasures of the industrial food system. We should make improving them the object of our politics. It's what we call democratic hedonism.

Providing people with accessible and abundant pleasures is a morally important task and a worthy basis for political action. It is not mere hedonism: Only when hedonism encounters democracy is it worth defending with any passion. Democracy demands a pluralistic and inclusive account of what a good life is, who gets to define it, and who is entitled to it. Making sure that everyone has abundant access to pleasure should be a social project that inspires collaboration, care, mutuality, and solidarity: for other people, for our common planet, and even for the animals with whom we share it.

When it comes to pleasure, the memory, context, and conviviality of food are every bit as important as its flavor. We have eaten more expertly prepared fine-dining meals than we care to remember, and, indeed, we don't remember many of them. But what do we remember? Gabriel remembers Chicago-style hot dogs at the Dog n Suds in Montague, Michigan, with his old man. Jan could live off pizza and espresso the rest of his life and can tell you exactly where he's had the best and with whom (shout out, all past members of the Wellington Pizza Club). We remember laughing together on that Brooklyn morning about just how much better the breakfast sandwich was than the previous night's dinner at one of the most expensive restaurants in the country. It was the meal that sparked the idea for this chapter. These food memories and preferences aren't reducible to only flavors or seasoning. Food doesn't work like that. We experience it as social creatures, and the pleasures we take from eating exist somewhere between the delicious morsel and the context or the people we enjoy it with—in the interstices between the sensuality of experience and the grander abstract scales described by ideology, theology, and social theory.

Much of what works with our food culture recognizes and cherishes those pleasures, even when they are expressed in distinctive aesthetic forms: Farmers' markets and greasy-spoon diners both offer people the pleasures of conviviality. Most people, Right and Left, rich and poor, want food that is healthy and fresh for themselves and their families. Beyond that, it's true that some people yearn for elite (and costly) experiences, but most normie pleasures are grounded in values that are tough to square with snobbishness: affordability; availability; familiar, beloved flavors, ingredients, and techniques; and convenience. What the snobs miss is that most people don't want eating to be a tiresome chore, another laborious task jammed into a day already as overstuffed as a leaking calzone.

Pleasure and choice in society are not equally distributed. Not everyone has the same menu, and some people can select only from limited and unappealing options. Social position, limited resources, energy, and time, and uneven knowledge shape access, and the result is that some pleasurable snacks are too high in the cupboard for some people to reach. And other people have handy stepladders that they're just not sharing. Meanwhile, people often do things they do not want to do because duty, care, responsibility, and necessity compel them. The gendered labor of food preparation, for example, means that common foodie clichés about the "pleasures of cooking" or avoiding "processed" foods can intensify sexist household burdens or, at the very least, be experienced very differently by men and women. Foodie writers often demand that Americans should eat more home-cooked meals. But the majority of the cooking labor in households, even in 2025, is likely to be done by women. Demands for daily from-scratch dishes and market runs will intensify the burden of what sociologist Arlie Russell Hochschild famously termed working women's unwaged "second shift."[19]

All this means that democratic hedonism requires that sensual pleasures be analyzed in relationship to some tempering values: capacity, access, and pluralism.

If we use the "capabilities approach" of philosopher Martha Nussbaum, examining *capacity* requires us to consider not only the pleasures of the moment but also a more holistic understanding of what makes a good life: the pleasures of a lifetime well lived. Capacity requires us to think not just about what we *are* but also about the full potential of what we may *become*. As Nussbaum notes, it is only when people have their basic needs met that they can achieve their full potential for thought, creativity, joy, love, and pleasure. Diets that are momentarily pleasurable but that make people ill in the long term diminish their future capacities over a lifetime and can burden their loved ones, friends, and communities. This is one of the reasons nutrition and health have to stay in the picture. We would not say, for example, that just because someone would find it pleasurable to eat an exclusive diet of Giordano's deep dish, they should do so. Doing so will probably ruin their health and destroy their capacity for future pleasure. Have a salad once in a while![20]

Similarly, following scholars and activists of disability, an emphasis on *access* requires that we consider how social structures, the built environment, and political power shape the menu from which we select. Sometimes we have to proactively reshape the world to ensure that everyone has adequate access to the things that make up a good life, building new infrastructures that expand access to people who would otherwise be left out. Expanding access to pleasures for some will require imposing costs on others. More funding for school lunch programs, for example, could require taxing pleasurable luxuries. Too bad!

Finally, we are mindful of the stubborn empirical reality of human diversity. "Axiom 1: People are different from each other," wrote literary critic Eve Kosofsky Sedgwick, and we'd all do well to heed it now. Because tastes differ, democratic hedonism requires *pluralism* about the pleasures of food, one more invested in empowering people to pursue their own pleasures as they define them instead of smugly dictating to other people what is (and is not) tasty. Like Jan, for example, you may be perfectly disgusted by Gabriel's desire, when his husband is out of town, to take a ball of mozzarella cheese from the refrigerator and eat it like an apple with his bare hands, but that feeling has little value in a debate about whether stores should stock mozzarella cheese and whether Gabriel should buy and consume them.[21]

Considered this way, democratic hedonism pushes us to ask how we can build a society of abundant and accessible pleasures. It requires suspending the belief that food and agriculture were once purer and more wholesome, and, therefore, that the pleasures people took from food were ever as innocent, pristine, and morally black-and-white as many food writers often imply that they both *were* and *should be*. Pleasure needn't exist only in imagined pasts and austere economies of scarcity. A central premise of this book is that whenever possible, we should try to offer people food rooted in economies of abundance. The good news is that the extraordinary productivity of American farming already makes abundant and affordable food pleasures possible. Further improving the American food system will require engaging and improving agriculture's productivity, not tossing it on the compost heap.

Chapter 2
FARMING WITHOUT SENTIMENTALITY

THERE MAY BE NO PLACE IN THE UNITED STATES WHERE FARMING packs a bigger economic, political, and cultural punch than Iowa. That makes it an ideal place to start when examining food availability in the United States and asking the big question of who produces what food and why.

People have been farming corn in Iowa for a very long time. The Native peoples of what archaeologists call the Northeastern Woodlands Culture grew maize in the rich bottom lands of the Mississippi's tributaries for centuries before contact with Europeans. Their agriculture and diet were, in the parlance of our time, mostly plant-based. They had no horses or cattle to pull plows, so they practiced no-till agriculture and rarely ate meat, which they hunted. It worked well. Modern agronomists estimate that their agriculture was likely more productive per person than European plow-based wheat agriculture of the same period and that it persisted for centuries without declines in productivity or soil fertility.[1]

In the nineteenth century the US federal government forcibly removed, with few exceptions, Native tribes from the state and sold their land to settlers. Those settlers immediately put Iowa's rich soil to similar use, but with some important twists. They plowed land and manured it with livestock. They grew corn for an emerging global market, not their own diets. They owned their land as individual proprietors.[2]

The resulting farms sometimes resembled the "family farm" ideal most Americans reach for when they think of farmers: a farm run by a pitchfork-bearing farmer Pa and his grim-faced farm wife Ma, those stoic sorts immortalized in Grant Wood's *American Gothic*. Wood was a native of rural Iowa, but when he painted *American Gothic*, in 1930, he was satirizing a bygone pastoral ideal that seemed to mostly cultivate stifling moralism, poverty, and hardship in its residents. (And the models for the painting were likely brother and sister, not husband and wife.) Nevertheless, *American Gothic*—or something like it—is now firmly lodged in the public's minds as a realistic depiction of how farming once was and, perhaps, should be again.[3]

But few of Iowa's classic small farms have survived to the present. Only 1,900 officially recognized "century farms," those that are operated by the descendants of the original settler families, remain. And many of those practice conventional, industrialized agriculture of the sort that dominates American farming and creates both an abundance of readily available food for consumers and a mess of social, environmental, and political problems.[4]

Farming covers 85 percent of Iowa's land, and most of the farms grow some combination of only three major crops: corn, soy, and pigs. Over 60 percent of all land in the state is dedicated to just corn and soybean fields, which produce an astounding 2.5 billion bushels of corn and 600 million bushels of soybeans every year. The corn lining Iowa's fields isn't the sweet corn you may see in your grocery, a variety that is packed with easily digestible sugars and that is delectably munched right off the cob. Iowa's corn has higher levels of starch that make it ideal for industrial processing but indigestible to humans. Pigs, cows, and chickens can still eat it, though. That means that over half of Iowa's corn goes to ethanol and most of the rest, along with most of the soy, goes to animal feed. There are around seven pigs for every human in the state of Iowa, and one in every three swine slaughtered in the United States comes from an Iowa farm. That's more than 14 billion pounds of pig per annum.[5]

And the state is paying the environmental price: What Michaelyn Mankel, an environmental activist based in Des Moines, described as a "water pollution crisis" driven by "sewerless animal cities" has contributed to the state's skyrocketing cancer rate—the second highest in the nation—and requires perpetual treatment and cleanup that costs Iowans about $66 million per year.[6] More than half of the state's waterways are too polluted for swimming or fishing, as are three-quarters of its lakes. However, discussions about those costs rarely make it out of the Iowa press and into the national media because, as they do with *American Gothic*, outsiders usually view Iowa agriculture through the lens of nostalgia. It's where presidential hopefuls make mandatory campaign stops at the State Fair to demonstrate everyman credentials over a plate of barbeque. It's a place that people talk about when they try to talk seriously about agricultural policy and American farming even if they've never set foot in the state. It's a place bathed in comforting fantasies about American farming: that farmers are hard done by and we need to do right by them, that we need to differentiate family farms from factory farms, and that farmers will be in the vanguard of saving the food system from the ills that afflict it.

But that's not the case. Instead, Iowa is where American agriculture is taken to its extreme. It has incredible productivity that is put to the wrong uses, and for the most part unchecked by regulation, it is left almost entirely to the market, leading to terrible outcomes for people and the planet. Understanding the business structures and incentives behind both the good and bad of its mass-scale, industrialized food system means starting with farmers. Most food writers hold farmers up as the key agents of progressive food system change, but the truth is a far less sentimental one: Farmers are businesspeople, and they use the tools they have to produce what they are incentivized to produce. That may be a tough pill to swallow for foodies who wish to make common cause with farmers, most of whom are not reformers but ardent defenders of the status quo.

We left sentimentality behind and headed to Iowa.

IOWA'S FECUNDITY WAS NOT EVIDENT IN EARLY MAY 2024 WHEN WE visited, although we did spot from a distance off winding country roads a few confinement sheds, the gargantuan structures where those millions of hogs spend their lives eating and pooping, packed flank to flank. Rainy weather had delayed the season's planting, and as a result, the slow rolling hills of east-central Iowa were still bare and dull brown. In just a few short weeks, those hills would be covered with green buds—corn and soy shoots—and the dull brown of the early spring would be transformed into a verdant pastoral.

In its own way, rural Iowa is a beautiful and prosperous place. The type you'd expect farmers to be proud of and gregarious about. And they are. Just not about business. Despite the centrality of farming to Iowa's landscape, economy, politics, and identity, many farmers are downright tight-lipped, at least with outsiders, about how they make their money. Maybe it's the Midwestern modesty. We managed to talk to a few, but here we'll focus on three who, in many respects, are archetypes that illustrate the broader structure of farming, not just in Iowa but throughout the USA as a whole. (We pseudonymized all three at their request.)

First, meet Ron. Ron is one of those "century farmers" and is every bit the stoic fellow of few words you'd expect to find standing with a pitchfork next to Ma. But his farm, a highly industrialized operation that grows corn and soy across thousands of acres, is a far cry from small-scale old-timey agriculture. He's what we call an Earner, a person who owns one of the top 5 percent of farms in the country in size and revenue.

Then there's David, a wealthy landowner who has never farmed but instead rents his land to people who do, which mostly means more corn and soy, as well as some pretty good income from the government "conservation" programs that pay farmers to leave marginal acres fallow. He's a member of what we call the Gentry, someone who owns and primarily derives passive income from farmland.

And then there's Tom, who has bucked the trend of industrial farming and practices small-scale, organic, free-range pig husbandry. A producer of widely acclaimed pork, he's the sort of farmer that foodies adore and the media come to if they want to talk to a farmer who does things right. But he's also navigating an unstable market for high-end products, which puts him—at least nominally—in the category we call the Noble Losers, those farmers who want to farm good products well but struggle with the harsh limits imposed by land prices, labor costs, and limited demand for expensive artisanal foodstuffs.

There is a tendency in talking about American food—from food books to political speeches to actual policy documents—to treat farmers as homogenous, but that's far from the case. Zooming in on each of the three farmers we spent time with in Iowa lets us explain key dynamics of the US food business. In terms of actually shaping the food on your plate, the Earners are the most influential—they grow most of the food that you actually eat—so we'll start with Ron.

THE FOOD BUSINESS IN THE UNITED STATES IS HIGHLY CONSOLIDATED, with a small number of big players controlling most links in the value chain, and that starts with farms. There are 1.9 million farms in the

United States and just over 86,000 in Iowa. But the top 2 percent of all farms operate on about 34 percent of all cropland, with the bottom 50 percent operating on only around 4 percent. About 16 percent of farm income is brought in by the 3.5 percent of farms that are not family owned. But close to 50 percent, or the vast bulk of money made by American farms, is generated by just 5 percent of what the USDA classifies as "family farms."[7] This latter group is the Earners, who, like Ron, mostly operate large and midsize farms and do well for themselves.[8] In fact, nearly all of the large farms normally decried in the media and food writing by critics as "corporate" and "factory farms" are family owned. They may contract with large corporate agribusiness (more on that in a second), but many are themselves large agribusinesses.

Ron's family has been farming in Iowa since the nineteenth century. For more than a hundred years, his family raised swine alongside corn and fresh produce. When confinement sheds began to appear, bringing with them fetid manure lagoons, they quit the pigs, and today Ron works within the placid rhythms of highly mechanized, conventional corn and soy cultivation. But that corn and soy probably wind up as feed for livestock, so he's technically still in the pig business. He buys the same inputs year in and out—hybrid GMO seeds and synthetic fertilizers, pesticides, and herbicides—from the same agribusiness corporations, and he plants and harvests his acres with the same machinery and implements, with a few repairs and upgrades as they come, and then he sells to another set of agribusiness companies year in and year out. Despite being objectively wealthy—the land he farms would sell for an eight-figure sum—much of the day-to-day on-farm labor Ron does himself.

At harvest or planting, when so much work needs to be done in so little time, Ron hires on additional laborers. The USDA distinguishes between people like Ron who operate a farm, whom they call farmers, and the people they hire to labor on their farms, whom they call farm or agricultural workers. It's a crucial distinction. Although the number of farmers (and farms) in the United States has been steadily declining for most of the last century, the total number of farmworkers has been creeping up over the last few decades, and today the Bureau of Labor Statistics and the USDA tally around 1.3 million farmworker jobs against just under 3.4 million farmers. We'll talk about farmworkers at greater length in a later chapter, but the important thing to know for now is that it is farmers, not farmworkers, who call the shots on farms. A family farm like Ron's doesn't need to employ many people to function, but family farms can also be gargantuan operations employing hundreds of people.[9]

The paragon of this in Ron's neck of the woods is Iowa Select Farms, owned by Jeff and Deb Hansen. In the early 1990s, the then-newlyweds started hog farming as a side project and a small operation. But when they saw the profitability of scaling pig farming via confinement housing, they jumped on it, and Iowa Select eventually grew into the fourth-largest pig producer in the country, churning out about five million pigs per year from operations located in over half of Iowa's ninety-nine counties. The Hansens pushed the factory-farming model despite opposition from neighbors, environmentalists, and even some state politicians, with Iowa Select becoming a feared economic and political powerhouse. When they're not jet-setting, they live in a luxury gated community just outside Des Moines. But Iowa Select remains a privately held company. A family farm, if you will.[10]

Ron's farm is neither as big nor as nefarious as Iowa Select, but he still runs an operation in the top 2 percent in the country by size. The business model of Earners like Ron, be they in the hog, corn, wheat, orange, tomato, or almond business, is low-margin, high-volume production. Take corn. It's traded in a global market as a fungible commodity, meaning that beyond broad, legally defined grades of quality and type, the particular characteristics of the corn don't matter: Any bushel of corn of a given grade sells for the same price as any other bushel. Practically, this means that any given corn farmer like Ron can't make any additional profits by growing "better" corn. Instead, he generates profits by producing corn at lower costs. Because corn has low per-unit profitability (low margins), making money usually means producing lots of it (high volume). Successful corn farmers have aggressively pursued capitalization (substituting labor with capital inputs such as machines and pesticides) and economies of scale (increasing the size of their operations). Those farms have grown larger over the past half century as they have gobbled up their less successful neighbors.

Capitalization and economies of scale are broad sector trends, but there is still variation in terms of scale and degree of capitalization among the Earners that depends on what kinds of crops they grow, labor availability, climatic conditions, water resources, and state and local regulations. Wheat farms in Montana can stretch over tens of thousands of acres and are vastly more mechanized than, say, few-hundred-acre tomato farms in Florida. This is partially because the nature of tomato harvest makes it difficult to mechanize but also because there is an abundance of relatively cheap migrant labor in Florida, a product of national and state labor and immigration policies.

Farmers are one link in the great agribusiness chain of the US food system, responsible for the bounty of food available to Americans. Coined by Harvard business professors John Davis and Ray Goldberg in 1957, the term "agribusiness" conveys the idea that modern agricultural production is more a chain of actors than just a solitary farmer running a farm. Agribusinesses, according to Davis and Goldberg, are any businesses involved in agricultural production and therefore include not only farmers but also anyone who provides farmers with seeds, inputs, debt instruments, and machinery as well as those who purchase agricultural products, process them, and sell them to consumers.[11]

When a farmer like Ron buys seed, fertilizer, or a tractor upgrade, he likely does so from one of the largest companies in the country, be it tractors and harvesters from John Deere & Company or fertilizer from CF Industries Holdings. You probably haven't heard of the second one, but its plant in Port Neal, Iowa, just south of Sioux City, produces several million tons of fertilizer per year. CF needs a whole lot of natural gas to make all that fertilizer, which means that BP, ExxonMobil, and ConocoPhillips are also agribusinesses. So too are many companies you wouldn't think are in the farming business. Nearly every major pharmaceutical and chemical company, including Bayer, DuPont, Dow, Eli Lilly, and Pfizer, maintains an agricultural division.

When Ron sells his corn, he does business, sometimes directly and sometimes through a middleman, with a grain wholesaler such as Archer-Daniels-Midland Company (ADM), which makes as much as $100 billion in revenue per year. ADM's boast to being "a global leader in both human and animal nutrition" is illustrative if creepy: The corn it buys from farmers like Ron mostly becomes either corn syrup, a ubiquitous sweetener found in many processed foods, or animal feed.[12] Before it arrives on your table as pork or a Coke, the corn has to pass through the hands of several other agribusinesses. Food manufacturers such as PepsiCo and General Mills purchase corn syrup and add it to their products, and meat companies such as Hormel, Seaboard, Tyson, and JBS slaughter and process animals fattened with corn feed before selling that meat to those same food companies as ingredients for things like Totino's pizza rolls (owned by General Mills) and microwavable El Monterey burritos (also General Mills). Or the meat of those animals will become whole cuts that, packaged in cellophane, land in the case at your grocery.

Or that corn might become ethanol. Around 35–40 percent of the US corn crop becomes ethanol, meaning that of the 90 million acres of corn cropland in the United States, an area about the size of Iowa—30 million acres—is used to grow corn destined to go in cars' gas tanks. The federal government subsidizes ethanol production through a number of programs, including the Biodiesel Mixture Excise Tax Credit, which provides gasoline companies $1 for every gallon of ethanol they mix into their fuel. That program alone costs about $3 billion per year, and the total subsidy is probably several billions more. These programs were launched to generate a sustainable and homegrown energy alternative to petroleum. But it takes significant energy to grow corn and still more to process it into ethanol. When those energy expenditures are included, corn ethanol doesn't offer much carbon savings, and it may actually be more carbon intensive than conventional petroleum gasoline.[13]

Today, it's evident that the major reason the government subsidizes corn ethanol is to ensure that agribusiness supply chains continue to be low risk and stable for everyone involved. Beyond the more basic promise of making sure farmers don't go bust, these policies guarantee that corn is produced in reliable abundance despite there being no market for it among eaters. Any farm plugged into this system enjoys a stable value chain that operates with clockwork-like logistical predictability and efficiency. That's why Ron's income is reliable.

Scholars of food geography and political economy highlight the uneven power relations in food value chains, dictated in large part by the size of different players and how much financial pressure or demand they can exert. Ron grows corn and soy because it's the easiest choice: It's what most farms around him grow and what buyers are looking for in Iowa. That makes farmers like him ready buyers for seeds and fertilizers and machinery for that specific form of farming. And it makes him a reliable partner for processors who mill his product and sell it on. In turn, their demand depends on buyers for corn and soy products and by-products, whether in food products, ethanol, or animal agriculture. Similarly, retailers can turn the fortunes of major processors and distributors by carrying or not carrying particular products, which can reverberate back through the value chain. McDonald's pulling fries off its menu would quite literally devastate the American potato industry in one swoop. Countless cafés and supermarkets stocking oat milk has incentivized farmers to switch crops to meet the new demand.

This is also why government policy has been central in shaping the incentives of agribusinesses and the Earners. For instance, the USDA offers crop insurance—to help farmers mitigate the risk of bad harvests—that favors row crops and large farms that have more crops. Because these crops can be de-risked, farmers have more incentive to grow them. If corn and soy get subsidies and apples don't, why grow apples? Farmers are ultimately in the business of making rational, income-maximizing decisions. This creates the incentive for them to support programs that create new markets for their products, such as the biofuel subsidies. Even if they wouldn't say this quite as explicitly, farmers love big government, or at least government checks, and big farmers love big government checks most of all. If demand for any given product were reduced—if there were rational decisions made about ethanol being a very inefficient use of land or if factory farming were constrained by a reduction in global demand for meat—incentives would change. Nothing's stopping Ron from, say, planting soy more suited for tofu. But no one has made him a worthwhile offer, and he hasn't exactly been looking for new opportunities.

The various forces pushing farms and agribusinesses to get bigger also push the industry toward oligopolies, when a few sellers control a market, and oligopsonies, when a market is dominated by just a few buyers. Take pork. The country's five largest producers, including Iowa Select, make up over 30 percent of total national production.[14] They sell to processors, where the four largest ones buy up and slaughter almost 70 percent of the nation's hogs.[15] And they sell mostly through retailers, a sector where four huge players control over 60 percent of the market.[16] The absence of a competitive marketplace permits big players, whether as buyers or sellers, to push everyone else around.

As journalist, activist, and farmer Austin Frerick has shown, similar dynamics prevail across many agricultural commodities, from pigs and chickens to corn and soy to berries and cucumbers. In these industries, actors he calls "Barons" emerge—the largest of the large and the most ruthless of the ruthless, like the Hansens and Iowa Select—who, de facto, shape the economics and politics of the industry as a whole.[17]

In pig and poultry farming, for instance, farmers do not usually own the animals they feed and house. Instead, individual farmers sign a contract with one of just a handful of multinational processors. The corporations provide the farmers with feed and other basic inputs, and they agree to pay the farmers a fixed sum for every fully grown animal the farmer returns to them for slaughter and processing. Thus, these farmers have a limited set of variables over which they have any influence: They're responsible for the sheds where the animals are housed, the land on which the sheds sit, and hiring workers to feed them. Even if a farmer found an independent source for inputs like feed or tried to go it alone and produce their own animals their own way and sell them in the open market, they would struggle to find buyers. This is what is meant by power exerted in value chains. Those contracting farmers may own millions of dollars of land and equipment, but they make a few dollars per animal and so are reduced to doing large processors' bidding. And sunk costs make switching to different products difficult. In fact, a grassroots organization called Transfarmation has been created to offer transition grants to farmers who want but can't afford to get out of the factory-farming business. One such operation, 1100 Farm, just north of Des Moines, proudly touts its transition from pigs to fungi. But the fact there are so few of these shows the power and profitability of the contract (and the scale of demand for bacon over mushrooms).[18]

Frerick and many others, including senators Cory Booker and Elizabeth Warren and a slew of economists in both academia and policy think tanks, sensibly advocate for strong antitrust measures that would make markets for both inputs and commodities more competitive. This is crucial to creating a fairer food system for farmers and consumers alike. Oligopolies and oligopsonies, much less monopolies and monopsonies (where only one seller or buyer dominate), are bad for the economy and bad for politics, driving up prices and stifling innovation. More competition would be better for consumers and might be better for farmers, who could branch out, sell in more open markets, and perhaps take more risks in what they choose to produce. Yet these policies are *not* endorsed by most of the major farming industry groups, and farmers themselves overwhelmingly support politicians who vigorously advocate for the erosion of antitrust regulations.

Why? Aren't farmers good ol' American capitalists, and don't capitalists believe competitive markets create incentives for efficiency and innovation? It's an empirical question, but our research leads us to believe that most Earners think, rightly or wrongly, that they have already learned to advantageously navigate agricultural consolidation and deregulation. When farmers can't adjust the sale price of what they produce, they can still find other ways to cut costs. The most serious costs imposed by their business models tend to be borne by the public in the form of environmental degradation and by their workers in the form of dangerous working conditions and low pay. They are, quite literally, invested in the consolidated system that the trustbusters rightly criticize. And why wouldn't they be? It's good money, and it rolls in like clockwork. The profits are private, but the risks are socialized.

RON IS ADMITTEDLY A BIT COMPLACENT ABOUT HIS BUSINESS MODEL. The May rain is forcing a late planting, and he's not really doing much of anything when we visit. In fact, once the corn and soy are in the ground, there's not much to do, either. Conventional corn and soy agriculture is a whole lot of waiting punctuated by a few big moments: planting, harvesting, and cashing the checks, year after year, with little incentive to do otherwise and much incentive to do the same. Ron's choice of crops is particularly low on demands for hired labor, with the work of plowing, sowing, and reaping almost entirely mechanized. But this, in itself, speaks to the incredible productivity of American agriculture. That we have such a surplus of staple crops—that we can use them so inefficiently, burning them for fuel and mulching them into animal feed, and still have a surplus of crops for humans to eat—is a phenomenon unheard of in human history until the last century. Modern agriculture, with its combination of technology and science, ensures that food availability is not a concern in the United States and, indeed, in many parts of the world. Coupled with international trade, it ensures a world of incredible abundance, variety, and convenience. In fact, the extremely high levels of productivity in the food system mean that most Americans simply don't have to worry about basic questions that haunted our ancestors such as "Will there be enough food to make it through the winter?"

One thing we can't do without is staple field crops of the sort Ron grows. Just three such crops—corn, rice, and wheat—account for half of all calories consumed daily around the globe. Growing conventional staples efficiently, reliably, and at large scale is almost certainly a good idea. And an old one. Many staple field crops have been predominantly grown throughout human history at larger scales, often in monocropped rows, usually for both trade and subsistence, and using a variety of mechanical implements.

The establishment of settled agriculture is what historians call the First Agricultural Revolution. Rich soil, flat land, and access to water make scale in agriculture both possible and desirable, and have been the base of great civilizations and commerce. Take the "black soil" in what is now Ukraine, rich in ammonia and phosphorus, which was at the center of a grain-trading network that spanned the entirety of Eurasia going back to prehistoric times.[19]

As early states dependent on settled grain cultivation learned, this mode of agriculture has strict—and sometimes brutal—ecological limits that can bring things to a crashing halt if not carefully observed. Over time, irrigated cropland becomes salinated and poisonous to crops. Land that is over-tilled or left without cover crops loses its fertility to the wind, just as cropping the same plot of land, year in and year out, depletes its fertility and leads to declining yields. Early nineteenth-century British economist Thomas Malthus, from whom we get many modern misconceptions about overpopulation, described a cycle of booms and busts, where populations would grow until they outstripped agricultural limits and then collapse. But Malthus couldn't anticipate the boom in agricultural productivity in Europe (and later the United States and then the world) between the eighteenth and twenty-first centuries.

This was the Second Agricultural Revolution, marked by better transportation, wiser land and labor management, and advances in agronomy in Great Britain and then in much of the world. English wheat yields more than doubled between 1700 and 1900, and the consistent use of crop rotation, manuring, and careful seed and livestock selection made the UK and other parts of Northern Europe the world's most productive agricultural lands of the era.[20]

But the incredible abundance of the modern food system puts all this to shame. It is the product of the "green" or "third" agricultural revolution of the twentieth century, often associated with the agronomist and Nobel laureate Norman Borlaug, who pioneered modern wheat crossbreeding. This step in the evolution of farming is summarized by science writer Charles Mann as a productivity trifecta made up of superefficient fertilization, irrigation, and genetics, a package of technologies that turned the world from one routinely struggling with food availability to one where—in the aggregate—we now produce enough calories and protein to feed every person on Earth.[21]

Take corn, Ron's primary crop. Whereas an average acre would yield 20 to 30 bushels of it in the nineteenth century, in 2023 a new record was set of 177 bushels per acre.[22] Nitrogen and potassium are plant food, and Ron's corn slurps it out of the soil with gusto. To keep his soil fertile and his corn well-fed, he uses conventional synthetic fertilizers he purchases from the likes of CF Industries Holdings or its Canadian rival, Nutrien. Those companies use an updated version of a process invented in 1910 by the German scientists Fritz Haber and Carl Bosch. The Haber-Bosch process "fixes" atmospheric nitrogen as liquid ammonia that can be easily spread on soil with an applicator implement that Ron hitches to his tractor. Turning atmospheric nitrogen into plant food also entails the use of a whole bunch of fossil fuels. The Haber-Bosch process catalyzes a reaction between nitrogen and hydrogen, and the hydrogen is usually extracted from natural gas. This, incidentally, is why modern agriculture is so hard to decarbonize: Petrochemical inputs can't just be swapped out for clean sources of energy.[23]

Genetic technologies are evident in the other cornerstone of the entire enterprise: corn seeds, bred to be amazingly productive and genetically modified to be resistant to herbicides. Genetically modified organisms (GMOs) have been a perennial boogeyman in food writing and public discourse since their invention, and certainly since mass protests in Europe in the 1990s cast them as a threat to human health. In the United States, GMO seeds are ubiquitous, but mostly in commodity crops like corn.

GM seeds, like GMOs more broadly, have genomes that have been altered to include sequences that do not appear in nature in the same plant. This is called recombinant DNA, or combining genetic materials from diverse sources.[24] In cases of "cisgenic" modification, scientists take genetic material from one organism and insert it into the genome of another organism of a closely related species, while "transgenic" modification involves transferring genetic material from an organism of one species to an organism of another. These techniques enable scientists to modify organisms much more quickly and with far greater precision than the conventional techniques that Borlaug and his colleagues used to breed hybrid seeds in the 1950s, although with the same basic logic: Scientists search for advantageous traits produced by natural selection and then use genetic modification, rather than selective breeding, to create species members that exhibit those traits. More recently, gene-editing technologies permit fine-tuned alteration in gene sequences without introducing any foreign DNA at all; instead, tools such as CRISPR can in effect rewrite the DNA an organism already has.

GMOs are foundational to the great productivity and simplicity of Ron's farm. His corn plants are bathed in glyphosate, a powerful herbicide that kills the plants and weeds that would otherwise crowd and compete with the corn for water, sun, and nutrients. But Ron's corn, a transgenic GMO, is unaffected by the herbicide because it has genetic material from bacteria spliced into its DNA. That makes it resistant to Roundup, the patented glyphosate herbicide sold by the same company that did the splicing, Monsanto, the agribusiness titan acquired by Bayer in 2018. Monsanto's Roundup Ready line of GM plants are planted across hundreds of thousands of acres in Iowa. But this often indiscriminate use also now makes glyphosate, a known toxin, a mainstay in the state's water supply. But glyphosate itself is not a GMO, and it is also widely used in places like Europe that don't use GMOs. The problem, in other words, is the known chemical and not the GMO. But in this particular case the use of the GMO incentivizes the use of the chemical.

Those harms bring into stark relief how unproductive debates about GMOs have been for decades. Many Americans avoid GMO foods because of widespread skepticism—if not outright fear—of eating unnatural "Frankenfoods" spread by GMO critics and the media.

But whatever legitimate safety concerns scientists and the public may have harbored about GMOs when the technology was first developed in the 1970s, the GM horse is well out of the barn. Decades on, there is no persuasive scientific evidence that they are hazardous to human health. Meanwhile, seemingly well-intentioned groups like the Non-GMO Project, a third-party labeling group that producers can pay to certify their products as non-GMO, may entrench popular fears of GMOs by suggesting that unlabeled products are somehow suspect. It's not just that GMO foods are safe to eat; it's also that the idea that they're unsafe to eat distracts people from the harms of some forms of commodity-crop production and the diversion of crops away from human food to animal feed and ethanol, where GMO use is ubiquitous and where its productivity is squandered on malignant ends.[25]

This is yet again a question of incentives and not technology itself: Ron's field is both a wonder of modern technology and a profound policy failure! Indeed, factory farms and biofuel are the result of decades of business pressure on agricultural-state politicians, who in turn push for lenient government policy. Agricultural historian Ariel Ron refers to this as a self-reinforcing and ultimately counterproductive "iron triangle" that makes farmers like Ron rich and diverts the gains of modern farming technology to ultimately irrational and environmentally harmful ends.[26]

But looking to agriculture outside Iowa, the same productivity has been applied to countless foods. There is no shortage of food calories produced in the United States, including an abundance of fresh and healthy fruits, vegetables, beans, pulses, and legumes, all grown using similar principles. The apples and cherries of Washington state, the pears of New York state, the peaches of Georgia, the tomatoes and oranges of Florida, the spinach and greens of California. And those productivity numbers for corn? They're similar for wheat, like that grown in the Dakotas. We can produce fruits and vegetables and grains like clockwork, with that trifecta of irrigation, fertilization, and seed technology driving productivity on farms around the country.

Making farming better will involve incentivizing farmers to grow different things and farm fewer acres overall: more food, less feed, no biofuel. But it will also involve improving agricultural technologies, including through broad public investment, to both further improve productivity and to reduce the sum total of hazardous chemical inputs used in farming.

Food scholar Robert Paarlberg notes that although total productivity in American agriculture has grown by around 250 percent since 1948, the use of inputs like fertilizer and pesticide per acre hasn't risen; in fact, it peaked in the 1970s in the United States. Rachel Carson published her searing critique of the insecticide DDT in 1962, leading to its ban in the United States in 1972. Pesticide use has declined 80 percent since then. We now produce more on less land and with fewer chemicals, and we haven't reached the limits of how much more productive and efficient we can get. Paarlberg's favorite bits of ag tech for further improving this come under the heading of what the industry calls "precision agriculture," the integration of modern computational technologies into farm operations.[27] These technologies, for instance, cut down on indiscriminate herbicide spraying through targeted "spot application," reducing impacts (and costs) without disrupting how most farms operate.

Some of our favored ag-tech products are seeds. As we were writing this chapter on Iowa, we attended a dinner organized during New York City's Climate Week by Triple Helix, an agricultural-technology nonprofit, and Rethink Food, an NGO that works to address food waste. Every item on the menu contained at least one ingredient that had been genetically modified, gene edited, or produced through a technological process like precision fermentation. There were Arctic apples from Washington, which use transgenesis to slow down or "silence" oxidation, staying crispy and white for hours even when peeled, thereby extending shelf life and reducing food waste. There was a polenta dish topped with Norfolk Healthy Produce purple tomatoes from California, purple because they have two genes implanted in them from snapdragon plants that give them a massive antioxidant boost, making them healthier and giving them longer shelf life as well. For dessert there were papaya bars made with Rainbow papayas from Hawaii, genetically engineered to resist the ringspot virus that once threatened the viability of the state's papaya industry. The meal was delicious and eye-opening.

And it's not just produce that can benefit from this technology. There are huge wins to be had with staple crops as well. The International Rice Research Institute, a public research institute funded primarily by the Philippine government and international philanthropic foundations, conducts research on a GM rice variety called Golden Rice that is genetically modified to have significant quantities of beta carotene, a precursor of vitamin A, which is otherwise absent in rice. Golden Rice received FDA approval for sale in the United States in 2018 and was approved in the Philippines in 2020. Of course, vitamin A deficiency is vanishingly rare in the United States, but in the developing world it causes complete or partial blindness in somewhere between 250,000 to 500,000 children every year. Tragically, Golden Rice has been targeted by anti-GMO activists—many of them, like Greenpeace, groups from developed countries—which have successfully stymied its launch in many of the countries that would benefit the most from it, including the Philippines.[28]

Today, one of the biggest barriers to the development of potentially transformative GMO and gene-edited products is the warrantless fear that the products are dangerous to eat. This is a bitter irony. In the name of holding agribusinesses in check, anti-GMO activism effectively helps agribusinesses like Bayer maintain seed-technology oligopolies in crops like corn while stifling the development of GMOs that could make for better food like that we tasted during Climate Week. The sooner we snap out of our distaste for genetic modification, the better. Meanwhile, we should address the incentives that undergird Ron's and Bayer's business model, including implementing penalties for glyphosate runoff under the Clean Water Act and scrapping requirements for ethanol in fuel.

Iowa is a state where all the incentives are all wrong. Farmers and innovators are pushed into growing crops mostly not eaten by humans, supporting an inhumane factory-farming industry, and despoiling the state's soil and water. It's easy to assume that this is an unavoidable state of affairs with modern Big Ag. And we agree that it's easy to get a bit jaded about farming when you read about how the Earners make their money, but we're the ones—all of us—who are paying them to do it. In a sense, we are collectively wasting an extraordinary asset, even as we lay the blame at the feet of bad actors.

DID YOU KNOW THAT MICROSOFT'S COFOUNDER BILL GATES IS THE nation's biggest individual owner of farmland? He owns around 260,000 acres of it.[29] This fact has drawn quite a bit of attention from the more unhinged corners of the internet, touted as evidence of a conspiracy of global elites intent on controlling the food system and making Americans, to quote right-wing provocateur Mike Cernovich, "eat bugs, live in a pod." But the truth is likely much more banal. Gates, it's safe to assume, is simply attending to the sage advice of Tony Soprano: "Buy land, AJ. God ain't making any more of it."[30]

And although Gates probably doesn't fit your mental image of a farmer, maybe he should. The median American farmer is pushing sixty; Gates is sixty-nine. About 95 percent of farmers are white, and around two-thirds are men. Gates is very similar to many American farmers in two other ways as well: He's a hell of a lot wealthier than other Americans, and he doesn't actually do any farming himself. This makes him part of what we call the Gentry. We didn't spend time with Gates in Iowa, but we did with another major landowner, David.[31]

A charming man with a wry and sardonic wit, David grew up in Iowa and joined the Air Force after college. When he left the service, he found himself with inherited acres spread over several counties near Des Moines. He is a shrewd businessman. He knows how much his land is worth and, roughly, how much income each acre should generate. But he leaves that work to people with muddier boots. He doesn't farm any of the land he owns, and he never has. Instead, he rents it to his neighbors to farm or lets the government subsidize him for leaving some of it fallow. He lives in a small town close to his daughters, their children, and his extended family—being a grandfather is his primary passion—and the combined income from his pension, savings, and farmland furnishes his whole family with a very pleasant standard of living. He drives a really nice car. He is well traveled and refined, with a fondness for fine steak and cigars. He is, if anything, a country gentleman.

Because he doesn't do any actual farming, the May rains don't bother him one way or another when we visit. After all, how they affect whoever is farming whatever they're farming on his property isn't really his problem. In fact, when we ask which company's seeds are planted on his land and which agribusiness the crop is sold to, he doesn't know. His lifestyle is a better match for the original meaning of the term "family farm." In the nineteenth century, a "family farm" tended to mean a large country estate owned by a wealthy person but usually operated by someone else, to which the owner might eventually retire but where they did not necessarily reside.

This arrangement had predecessors in Old World agriculture. Prior to the nineteenth century, European royals, like Bill Gates today, were frequently the biggest owners of farmland, and many members of the European gentry (small "g") like aristocrats and nobles held their wealth in farmland and depended on rent income from that land. Few people thought of the king and his court as "farmers," of course. Sure, some of the European gentry did express varying degrees of interest in husbandry and agriculture, often as an art or a hobby, but most were more likely to focus their energies on the great pastimes of the rich— leisure, politics, and war—and allow the peasants to handle the drudgery of farming. This was precisely why the possibility of land ownership seemed revolutionary and attractive to many European immigrants to the United States.[32]

We offer this brief bit of historical trivia because it provides a useful perspective on farming: It's not any particular form of labor, but rather a property relation to land. Although some farmers certainly work hard on their farm, that doesn't make them farmworkers, the laborers paid a wage for their work. Being a farmer means being a landowner. And this helps make sense of the Gentry, who own farmland but don't farm. Before you stop us to protest that surely this is some fringe group, consider that in Iowa, as agricultural a state as agricultural states get, 60 percent of farmland owners don't farm and . . . wait for it . . . about a third have no direct farming experience whatsoever. In Iowa, folks like David are the *majority* of farmers.[33]

So what's going on?

Farmland is a prized, lucrative, and steadily appreciating asset, one many people would like to own. An average acre of agricultural real estate in the United States is now valued at around $4,000, up from $2,500 per acre only a decade ago, a figure that has been rising, more or less steadily, for the last thirty years. For context, an average American farm, at roughly 450 acres, is worth about $1.8 million on the open market, and Gates's haul is worth over a cool billion. Cropland, the subset of farmland used specifically to grow crops, is even more valuable. In some places, such as California and New Jersey, cropland can bring around $15,000 an acre. In Iowa, good farmland goes for well north of $10,000 per acre.[34]

This reality is increasingly drawing not just wealthy investors like Bill Gates but also institutional investors. One of the biggest investors in Iowa farmland is, of all things, the Church of Jesus Christ of Latter Day Saints. Through its investment branch, the Deseret Trust Company, it has bought tens of thousands of acres in Iowa that it leases out for corn and soy. These sorts of investments, in turn, drive up land prices, both disincentivizing small-scale farming and incentivizing land grabs by more investors looking to join the Gentry or those Earners looking to expand their operations while they can still afford it.[35]

Most American farmers are wealthy. Not just comfortably middle class, mind you, but *wealthy*. Not Gates-level wealthy, sure, but the median farm household in the United States has around $1.4 million in assets, against the national household median of $167,000. An astounding 98 percent of farm households have household wealth above the national median. Wealth, of course, is not income, but farmers are also doing well on that front, and they are disproportionately located in rural communities with lower costs of living.[36]

Meanwhile, farms are highly solvent enterprises, more so than the average American consumer and most small businesses. In 2023, on average, farmers owed just over 13 cents in debt for every dollar of their assets, a healthy debt-to-asset ratio. For all you may have heard about the woes of American farmers, in 2022 there were exactly 169 farm bankruptcies across all of America out of just under 2 million farms (and the number has never exceeded 723 in any year in the past quarter century). That's a bankruptcy rate of about 0.008 of a percentage point.[37]

The USDA tracks something called "limited-resource farm households," meant to be a good approximation of poverty, and these are the farms that news stories and foodie writers focus on when they bemoan the apparent plight of farmers in America. Only about 7 percent of all farm households qualify, and even that small group includes some farms that generate little income not because they are economically distressed but because their operators have retired from active farming.

It is true that many people who own farms derive most of their income from other jobs, but that's not necessarily a sign of poverty. The USDA classifies around 40 percent of all farms as "off occupation," meaning that the farm is not the operator's primary source of income. They also do well, earning, on average, around $160,000 per year and holding average wealth of about $1.8 million. Remarkably, average members of the Gentry residing on "off-occupation farms" report net losses of nearly $5,000 from their farming activities.[38]

In Iowa, over 80 percent of all farmland is owned debt-free and held for long periods of time.[39] The Gentry, like the Old World gentry, are sitting on an appreciating asset. It doesn't matter to them if, like half of all US farms, they do less than $10,000 in annual sales. However, they do access the public programs that support farmers. Remember the publicly subsidized crop insurance that staves off price volatility and ensures the stability of the Earners' agribusiness model? Only around 13 percent of all farmers make use of it. But the Gentry disproportionately access another form of public subsidies: Retirement and off-occupation farms take in, respectively, 25 percent and 29 percent of all Conservation Reserve Program payments, the USDA fund that pays farmers to retire land from active agricultural production to preserve biodiversity. It's a steady income stream for not doing much more than simply owning land.[40]

In the conventional narrative, small farmers would prefer to derive their income from farming, and the fact that 97 percent of off-occupation farmers earn a majority of their income from nonfarm employment is evidence of their financial precarity. Michael Pollan's *Omnivore's Dilemma* showcased one farmer, George Naylor, lamenting at length that it was a serious indictment of government policy that his wife's employment in town supplemented his farming income. Pollan offered Naylor as a stand-in for small farmers everywhere who wanted to make money in farming but simply couldn't. But if so many people want to make a living as farmers, why do so many farms, and the majority of small farms, do almost no business whatsoever? And why don't they access the public programs that would help support the business of agricultural production? Why, instead, would small farmers disproportionately access programs that pay them not to farm?[41]

If you start from a more evidence-based perspective, this apparent paradox dissolves. "These aren't the farms of the poor; they're the yards of the upper-middle-class," explained journalist Maggie Koerth.[42] For instance, David owns farmland as a real-estate investment vehicle much as you might own a house. Not only are the Gentry not competing with (nor are they losing to) the Earners, but they have some shared interests that keep them in alliance. Namely, the expansion and consolidation of large-scale farms creates demand for farmland that translates into higher land values, so regulatory policies that benefit the Earners by reducing costs and facilitating higher volume production also wind up benefiting the Gentry. There are, to be sure, also bound to be some conflicts—the stink wafting off a manure lagoon will surely spoil the serenity of your country estate, or an institutional investor might scoop up some prime retirement parcel—but on the balance, the Gentry tend to share political preferences with the Earners. Consequently, they are vital if often passive political allies of American agribusinesses. And their relationship to farming—land ownership—is a difficult one to challenge.

CURVEBALL: TOM IS NOT A LOSER AT ALL, NOBLE OR OTHERWISE. He's doing quite well for himself, thanks for asking. But his path was narrow and rocky, navigated with tenacity, skill, and a helping of luck.

Most people who try to blaze similar trails, well-intentioned and driven by idealism and passion, wind up at the bottom of the ravine, members of what we call the Noble Losers, farmers who adopt niche and artisanal production modes because of ideological commitments. In business-speak, they are "mission driven."

Sensitive and thoughtful, with hair that is just starting to gray, Tom is a local boy made good. Two decades back, he returned home from college inspired to take a different path from the pork factory farming that he viewed as environmentally destructive and cruel to animals. As companies like Iowa Select industrialized and befouled rural Iowa, Tom took his small piece of land close to Cedar Rapids, just a short drive from the iconic baseball field carved from cornfields in the Kevin Costner blockbuster *Field of Dreams*, and deindustrialized.

Tom's swine love the late spring rains that are proving a nuisance for corn and soy farmers like Ron. Pigs sunburn easily, which isn't a concern for the factory farmers whose pigs are confined indoors for all of their lives and never get to see the Iowa sky. But for Tom's free-ranging hogs, the clouds ensure that, between snoozes in the mobile huts that dot his fields, they wander through open pastures, munch on roughage, root under bushes, and enthusiastically roll around in the mud. Swine dung is watery and difficult to collect and spread, which is precisely why factory farms pipe it from confined sheds into noxious open-air lagoons. Tom's pigs, by contrast, do the collecting and spreading for Tom on their own four trotters.

It's an expensive way to produce a pork chop. Initially, Tom sold his meat—by reputation some of the highest-quality pork in the country—through fine-dining restaurants in places like Madison, Wisconsin, and Minneapolis, Minnesota. But reliance on them proved unprofitable and unpleasant. Orders were unreliable in timing and quantity, and Tom discovered that some restaurant managers would buy his products just once and then leave his farm's name on the menu indefinitely. There was—and is—no regulatory oversight of restaurant menus. He pulled his pork from most of the restaurants that had once ballyhooed it and now sells through only a handful of restaurants around the country whose owners he knows personally, instead focusing mostly on high-end grocery stores and direct-to-consumer sales over the internet. He caters to an affluent customer base that stretches to California and New York and, recently, as far as Japan.

Unlike many others who pursue this path, Tom makes it work. He is a charismatic businessman who succeeded because he actively built a consumer base that he scrupulously maintains through personal connections and personalized service. Although he's not vocal about it, he's flexible when it comes to integrating modern technology where it fits; he artificially inseminates his sows using semen sourced from a global sales network, for instance, even though that way of breeding swine has been common in Iowa only since the 1990s and, from an animal-welfare perspective, is highly dubious. That Tom can stay the course is a testament to how well he avoids mission creep—his mission is free range and high quality, *not* local or religiously low tech or even entirely ethical per se. And it's neither scalable nor affordable for most Americans. Tom's approach results in six-ounce pork chops that run just under $6, around three times the going rate at Walmart. So much food writing, including that of Pollan and his free-range farming guru Joel Salatin, implies that production processes that culminate in a $6 pork chop can and should be the norm. The reality of basic economics says they can't. That's why farmers like Tom are rare.

Many broadly construed progressive policies of the sort we favor in this book would help Tom and farmers like him. More vigorous antitrust and environmental regulations would put Tom on a slightly more level playing field with the folks at Iowa Select who shove their swine into concentrated animal feeding operations (CAFOs). That may be why Tom's politics lean left and why he usually supports Democratic candidates, unlike David and Ron, who do not. But that preference is far from uniform for small farmers. Although progressive policy reform would probably benefit many of them, people so highly motivated by ideological preferences—the sorts of people drawn to mission-driven farming—are bound to have idiosyncratic and sometimes extreme political beliefs. For every Tom out there, there's a Salatin, who is convinced that any regulation, no matter how prudent, is tyranny, even if that sort of approach benefits the megafarms much more than the small guys.

Regardless, even if there were extensive antitrust intervention in the food system, Iowa can hardly return to a state of only small farms any more than it can return to precolonial agriculture. Food systems need to match the society they feed. To the extent that small, bespoke farms can play a part in the food system, it would require a steady customer base willing to consistently pay premium prices and policies that would try to make room for a variety of farmers to try different approaches and try to build out new markets and value chains. But that would require the availability of more cheap land.

For many smaller farms, the reasons for economic distress are ultimately similar to what you might find in other industries. It could just be bad luck, such as a vital employee leaving the farm at an inopportune moment. It could also be mismanagement. That might include technical mistakes such as applying too little fertilizer or leaving the wrong field fallow for a season. It might also include business errors such as purchasing unnecessary mechanical implements, taking on too much debt at unfavorable rates, or expanding more rapidly than the available pool of labor can support.

But others may be farming for noble reasons that are not compatible with turning an easy profit. No matter how well they manage their farms, they will eventually fail because they are producing at a cost that they have no reasonable hope to recoup from consumers. They may want to provide fresh, high-quality food to underserved communities, farm their acres in ways that are ecologically regenerative, or raise their pigs and chickens according to laudable welfare standards. Farmers in this category have their hearts in the right places, but they often lack a realistic business plan. Many of them are destined to fail, which is why we call this group the Noble Losers.

Some Noble Losers have bought into the foodie line and, as a result, have overestimated the public's appetite for niche and artisanal modes of agricultural production. Despite the enthusiasm of food writers, the market penetration for local and organic farming is not only tiny but has actually decreased as a percentage of total American food consumption over the past quarter century. Beyond incurring additional costs that frequently accompany rarified modes of production, "farm-to-table" business models pose serious obstacles well beyond the field and meadow. For starters, to succeed, the Noble Losers must also distribute and market their products, something that the Earners need never do. The most technically adept regenerative farmer will go bust if she doesn't know how to effectively market her products. Similarly, grocery and restaurant suppliers, for understandable business reasons of their own, usually decline to purchase from farms that cannot deliver a substantial volume at a standard quality on a reliable schedule, a practice that locks out many starting small producers from the reliable customers to whom their more conventional and established competitors sell.

Meanwhile, *local* farming must, by definition, be close to its customers, which subtly stacks the deck against success. The bad news that many starting Noble Losers discover is that the closer farmland is to consumers—and especially the deep pockets of consumers interested in and able to afford hyperlocal food—the more expensive it is. Some farmland is more expensive because it is better for farming—this is true in Iowa—but the most valuable farmland is expensive because it is located in places where there is high demand for land in general. A patch of land in western North Dakota may be optimal for farming wheat but is unlikely to be in demand to build a suburban subdivision. Not so in central New Jersey, California's Sonoma County, or the Hudson River Valley north of New York City. In these places, Noble Losers are competing for land against real estate developers.

Worse still, many of the potential farms a Noble Loser might like to operate are also prime candidates for the Gentry. Recall that many of the Gentry find farm living to be an attractive lifestyle but that they often derive their income from other forms of employment. For these folks, the proximity of a farm to a city means they can have their hog and eat it too, living on a farm but enjoying the amenities and economic opportunities that come with living near a city. They'll bid up farmland, which means that Noble Losers must compete with the Gentry for land even as their products must compete with the Earners for shelf space and market share. Or they must rent from them and make the money work. Finally, where farmland is pricier, the general cost of living is also higher, and labor is more expensive. Rarified niche and artisanal modes of production are frequently more labor intensive, and that means that Noble Losers must locate their operations in places where their largest costs—land and labor—are likely to be more expensive. Not good! And not easy to fix.

And it's not just tiny and artisanal farmers who learn that lesson the hard way. The same can happen to highly capitalized firms trying to "disrupt" farming. Take AeroFarms, founded in New Jersey in 2004, an industry leader in "vertical farming," a method for growing plants indoors through densely stacked trays irrigated through hydroponic or aeroponic technology. The theory was an urban farm with California-like growing conditions year round that provided local produce to New York City. But theory never quite translated to practice. Despite raising hundreds of millions of dollars in funding over two decades, AeroFarms never managed to find profitability growing and selling its major crop, lettuce microgreens. When it emerged from Chapter

11 in 2023, it downsized to just one farm, a facility in Danville, Virginia. The high-tech, high-capex approach of AeroFarms might seem a universe away from other Noble Losers looking to grow microgreens in regenerated soil in the Hudson Valley or free-range pork in Iowa, but in both cases being mission driven can blind farmers to basic empirical problems. It makes more sense to do vertical farming in rural eastern Virginia where real estate is cheap. But if that's the case, why not just convince a conventional farmer like Ron to grow microgreens? In fact, there would be many benefits for the planet and the soil if Earners like Ron shifted away from corn, soybeans, and livestock and toward fruits and vegetables. Infatuation with vertical farming among urbanists, investors, and tech elites is the funhouse-mirror image of the foodies' nostalgic agrarianism: a fantastical and doomed desire to escape the conventional food system rather than just improving the farms we actually have.[43]

MUCH FOOD WRITING AND THE AMERICAN REGULATORY STATE SHARE AN unfortunate trait: They are both riven with "agricultural exceptionalism," the tendency to see and treat agriculture differently from all other economic activities. The clichéd presentation of farming as "a cherished way of life" and "national heritage" gestures toward lifestyle and identity to conceal what are often simple economic motives. Farmers may have strong cultural preferences that lead them into farming, and they are drawn heavily from rural communities with distinct values and histories that are different in important ways from cities and suburbs. But most of that doesn't drive practical business decisions and tends to obscure what farmers share with their urban counterparts: a concern for the bottom line.

The picture we've painted of American farming and farmers is not as pretty as a Grant Wood painting. We can't tell you a compelling story of a hardscrabble farmer seeking to break rules as some sort of inspirational ideal (in fact, the one prototypically inspirational farmer we met seemed to want *more* rules when it comes to how people treat pigs, the water, and the land). It's not that the many farmers we spoke to, including Ron, David, and Tom, came across as selfish, reckless, or glib about their business decisions. They cared about their families, their communities, and the state of Iowa. But they worried about the profitability of their farms. And that's the rub. Farming is a business, for good and ill.

Much farming in America is amazingly productive and furnishes us with a bounty of safe and affordable food, including many healthy, nutritious options. This must remain the cornerstone of any food system. But, unchecked and unguided, this productivity can create perverse incentives for crops that no human will eat and for wanton environmental despoliation. Iowa is a prime example of this, its potential as a breadbasket corrupted by the appeal of easy money in biofuels and factory farms. It's a cautionary tale.

Transforming the food system by altering the behavior of farmers faces some serious constraints, but there are promising avenues for change. The simplest way to think of this is that appealing to the good intentions of farmers without appealing to their pocketbooks is unlikely to accomplish much of value, and making farmers do things they don't want to will require legislation.

If we want American farmers to sell more fresh fruits and vegetables to Americans, more Americans will need to buy more American-grown fruit and veggies. That might involve tools like incentivizing healthy food purchases in schools or through programs like SNAP. It could mean extending programs like crop insurance and other subsidies beyond corn, soy, and wheat. It might help to introduce value-added taxes that would make fresh produce less expensive relative to processed foods. But ultimately it means creating demand that farmers can supply at a profit.

If we want farmers to protect biodiversity more, we will need to incentivize them financially to do so. A range of scholars, businesspeople, and policy thinkers have argued for initiatives like a reforestation fund that would pay farmers for the CO_2 they would capture on their land if it were taken out of production and left to regenerate or reforest. This is referred to as land sparing, and it creates a carbon sink, cleaner waterways, and a healthier habitat for wild species. Its environmental benefits have been modeled to be so great that we'd be better off polluting more on fewer acres of land if the freed acres wound up as unmanaged forest or grassland. Some sort of land bank is a dreamy solution (perhaps too dreamy). A more promising idea is to expand existing programs that pay farmers to conserve their land and provide subsidies to farmers who reforest their fields.

If we want more small farmers, we will need policies that provide favorable loans or land-redistribution programs that allow younger and historically excluded populations to start farms. But we need to be careful, given that encouraging new farmers to try to make money from their good intentions alone is more likely to generate farm foreclosures than it is to cultivate a more just and accessible food system.

New outcomes will require voters, eaters, and policymakers to stop fetishizing farmers and to treat them like bad businesspeople when they do badly. Hog and pesticide runoff in groundwater is not an impossible problem to solve. Ban some things. Implement some fines. It may be difficult to muster the political capital needed to do this, but problems like these have clear-cut and reliable solutions. Sticks, not carrots, if you will. Farmers, of course, will protest that this will increase their costs. Yes. Exactly. The best way to disincentivize bad practices is to make them more costly.

You may assume from our defense of much of modern agriculture's fundamental soundness—especially when it comes to plants—that we favor only timid incrementalism when it comes to food production. Not quite. The most efficient technology we have for feeding people is photosynthesis—plants turning sunlight into edible calories and nutrients—and improving this through technology can only make it better. But there's one category of food that is inherently inefficient: meat. It's at once a source of grievous social, ethical, and environmental harms and one of the most cherished parts of the American diet. Can we find a way to preserve the pleasures of meat while dramatically reducing the costs of its consumption? We went to California to find out.

Chapter 3

IT'S THE COW, NOT THE HOW

O N A CLEAR DAY, YOU CAN STAND ON THE HILLS NORTH OF THE Abbot's Lagoon trailhead in Point Reyes, California, and take in a view as close to postcard-perfect American pastoral idyll as you'll find anywhere. Just across rolling green pastures, nestled in a small dip, are the squat buildings of a quaint dairy farm. The sun glints off its metal roofs, cows mill aimlessly in paddocks, and behind them the deep azure of the Pacific Ocean stretches to the horizon.

It's such a picture-perfect scene that it does, indeed, appear on postcards in souvenir shops near Point Reyes National Seashore. Similar images also bedeck the labels of the cheeses made from the milk of those milling cows, cheeses that appear on charcuterie boards at Sonoma County wineries and in upscale groceries in San Francisco, an hour south along the coast. This part of California is a hub of foodie culture. Here, local cheese pairs with the local wine, and the Patagonia and Carhartt sets clink glasses to talk terroir. The food chat around here, far from the industrial dairy megafarms you can find in California's dusty Central Valley, is about how going small, local, and regenerative can be a real sustainable alternative to the mainstream food system. And with the sun beaming down and your cheeks tingling from the sea breeze, it's easy to get caught up in the idyll.

California, unbeknownst to most Americans, has parlayed its temperate climate, rich soil, and (at least until recently) abundant water supplies into being by far the biggest agricultural producer in the United States, beating out heartland agricultural states like Iowa by tens of billions of dollars in annual revenue. But just like in the Midwest, Californian farmers have embraced the large-scale, industrial farming models that dominate agriculture across the country. The 20,000 square miles of the Central Valley are the state's agricultural hotbed. The astonishing productivity and scale of animal agriculture there have made California the country's single largest producer of beef, broiler chickens, eggs, turkey, sheep, and goats. California dairy products alone generated $10.4 billion in revenue in 2022, with the 1.7 million milk cows in the state producing the majority share of the nation's milk.[1] And with this scale come the same problems that plague industrialized animal farms everywhere else in the country. It's these farms you hear about when newspapers cover California's water overuse, environmental degradation, methane emissions, mistreatment of seasonal laborers, animal abuse, and most recently the spread of H5N1 avian flu.

Point Reyes is meant to be the sustainable foil to places like the Central Valley, a place where food production exists in harmony with nature rather than running roughshod over environmental limits. That's the story that first brought us to that awe-inspiring lookout spot.

But go stand in that exact spot on a day when the thick Pacific fog enshrouds the coast, blanketing the hills, obscuring the ocean, and turning the fenced-off cow pastures into muddy scars on the landscape, and you get a bit more contemplative. We went there accompanied by a local investigative reporter and several residents of the area who are fighting over land claims with local dairymen and ranchers, having read reams of reports about the area's dubious environmental record. With these on our minds, our clothes sodden from the damp, and our boots mud-caked, we wondered what else was hiding—or being hidden—in California's celebrated redoubt of small-scale farming.

You see, for all its apparently natural beauty, and despite its status as a National Seashore, there is little that is natural about Point Reyes. The rolling meadows and cow paddocks we saw were all protected by fences not just to keep the cattle in but also to keep native species of plants and animals out. Those native animals, like tule elk, frequently die from starvation or dehydration because they cannot access food or water that the farmers keep gated away exclusively for their cows. Those cows also produce more than 130 million pounds of manure every year, which leaches into the soil and groundwater and out into the ocean, choking local butterflies, songbirds, and salmon in effluent runoff.[2]

It's certainly not the so-called regenerative ideal of cows happily treading and fertilizing soil in some sort of cyclical natural balance. In fact, the problem with buzzwords like "regenerative" and "agroecological" is that they're notoriously slippery, with studies of the terms routinely finding dozens of different and often mutually exclusive definitions and no binding regulatory definitions like those that apply to terms like "organic." Rather than buying into buzzwords, understanding the environmental impacts of different food production systems and comparing them leads environmental scientists to use a basic set of metrics: land use, water use, greenhouse-gas emissions, eutrophication (water pollution), and harms to biodiversity.

Across all of these, Point Reyes is far from benign. It's not that the total environmental impact of animal agriculture at Point Reyes is larger than a comparable area in the Central Valley; it's far lower. But the reason *why* is the crux of the issue. A Point Reyes cow likely has a *larger* per capita environmental footprint than a factory-farmed cow, and commercial operations in the area have the same set of negative environmental effects as other dairy and beef operations. The aggregate impact in Point Reyes is not lower than conventional systems because of how its cows are produced; it's lower simply because it produces far fewer cows.

This teaches a vital object lesson about the environmental impact of food: It is *what* is produced, not *how* it is produced, that matters most. Beef and dairy production are inherently highly harmful to the environment. Smaller-scale versions of highly harmful types of food are the worst of both worlds as systemic food system solutions: inherently neither sustainable nor scalable. Common foodie solutions that bet the farm on shrinking the scale of production—going small, local, and regenerative—pale in comparison to solutions that would positively alter the composition of the average diet. But the lesson goes the other way too: Taking large-scale production of environmentally destructive foods and making it marginally less bad is far worse than taking a more sustainable product and scaling it up.

Farms like those at Point Reyes wind up selling what we call minced meat denialism. They downplay the intrinsic harms of meaty and cheesy diets, shifting emphasis to aesthetics and storytelling: dressing up the *how* and absolving the chow. The upside of this idea is that individual consumers won't need to make major dietary changes to address meat's costs—at least nothing beyond a slightly pricey shift from bad factory-farmed cuts to good meat from places like Point Reyes. But the theory of change here is convoluted and backward. Meat raised under (always vaguely defined) "good" conditions is no less harmful to the environment per pound than industrially farmed meat, but it is vastly more expensive, which inherently makes it a niche product. Absent a commitment to *eating less meat overall*, even those average consumers who could afford the odd indulgence in fancy meat would likely fill their plates with conventionally produced cheap meat anyway on other days.

That said, having clarity about the need to reduce meat's harms isn't the same thing as empowering people to alter their diets en masse. Leaving things at *eat less meat, stupid,* is what we call meat austerity. It presents reducing meat consumption as a virtuous sacrifice (maybe it is) that everyone should already be happy to make (they ain't). If social pressure and government policy must be directed to carve pleasures from diets, it should be done not with an unsympathetic shrug but with a robust investment in expanding pleasurable substitutes.

Meat is, in this regard, not unlike other bad pleasures you may also be loath to lose—booze, risky sex, dangerous sports, gambling, the list goes on. They all share this: You can often change what you desire or at least bend it into a less destructive course for yourself, your family, your community, and your society, but it takes hard, concerted effort and support. You need *more* options for pleasure, not *fewer*, as well as affirming and welcoming new norms, institutions, and products.

So how can we get there?

WE'D MEANT TO BEAT SAN FRANCISCO RUSH HOUR TO ARRIVE IN POINT Reyes in the morning, but Gabriel had clumsily locked us out of our Airbnb in Noe Valley, setting off a cascading series of delays and scheduling problems for our trip north of the city. By the time we crossed the Golden Gate Bridge, it was after noon, and Jan was hungry. Between Sausalito and Strawberry, he spotted a Panda Express and decided to indulge one of his great pleasures: orange chicken. Orange chicken is a fried nugget, crispy, salty, greasy, doused in a sweet and syrupy orange sauce that typically has just enough citrusy tart to distinguish it from its spicier cousin, General Tso's chicken. Both are favored, comically inauthentic bites of Chinese takeout for millions of Americans. Thanks to Panda Express's recent launch of an orange chicken made with Beyond Meat's plant-based chicken, it was a craving that Jan, who does not eat meat, could indulge right there in the passenger seat of the rental car.

Here's the remarkable thing about take-out orange chicken, a quintessential meaty delight: The chicken itself is mostly irrelevant. Most take-out orange chicken is made using tiny nuggets of low-quality, highly processed industrial bird, fillers, and stabilizers. It's the batter, oil, and sauce that make the dish, which is precisely why we'd wager many of our readers, if they were served Beyond orange chicken without being told what it was, wouldn't realize it was made from plants. But people like to eat the meat version. And that's a problem.

Meaty meals like orange chicken, beef burgers on your backyard grill, bacon at breakfast, and even the refined filet at a Michelin-starred restaurant *all* have dire environmental and social impacts. It doesn't matter whether they're fast or slow, junk or haute. Like we said: It's what is produced that matters most. *It's the chow, not the how.*

Americans haven't always eaten like this. The scale of the environmental damage of Americans' meat-saturated diet, as well as the cramped vision of dining pleasure that propels it, is a relatively recent phenomenon, caused in large part by modern animal agriculture completely remaking our diets and landscapes.

To really take in the magnitude of the resources required for animal agriculture, you need to zoom out to the planetary scale. That's exactly what the artist Mishka Henner was doing in 2012 when he was scrolling through Google Earth looking to map the extent of oil and gas operations in Texas. Instead, he found cows. Hundreds of thousands of them on the state's sprawling feedlots. And adjacent to all those cattle pens he found something else: the gigantic, effervescent lagoons of the manure produced by all those cows. The scale and contrast of the photos he took seem surreal: almost too big and too toxic to be true. The series, simply titled "Feedlots," has become one the artist's signature works, even landing him the coveted cover of *Artforum*, the most important contemporary-art magazine in the world. It's also a stunning visual testament to the size of the problem of livestock farming.

Agriculture takes up just under half of the world's habitable land. Land used to graze and grow feed crops for animals takes up three-quarters of that. Globally, we use a land area almost as big as South and North America combined only for meat. That leaves one quarter of all agricultural land—about the size of China—to grow all other food consumed by the 8 billion humans on Earth. In the United States alone, about 40 percent of total land area is used for meat: 650 million acres for pasture and 125 million for livestock feed.[3]

Land used for animal agriculture cannot be used for carrots. Or for condos. Or, perhaps most importantly, it cannot be left to flourish as wilderness. Each use of land, whether wild, urban concrete and steel, or bucolic farmscape, has environmental implications. Consider the Amazon. Chopping down dense thickets of trees so that cattle can munch on grass or so that soybeans can be grown to be fed to factory-farmed chickens permanently destroys the habitat of countless species of animals, changes everything from birds' migratory patterns to rates of precipitation, and turns the deforested land from a carbon sink into a source of greenhouse-gas emissions, primarily in the form of methane belched by cattle. This last trade-off is referred to as the carbon opportunity cost of land use: how much carbon would have been sequestered if land had been left to native plants.

Then there are the direct greenhouse-gas emissions. Food production makes up about 30 percent of all global emissions. Animal agriculture represents half of that, and cattle make up the biggest share.[4] The reason for the scale of these impacts is the scale at which we produce and consume animals. Globally, 92.2 billion land animals are killed for food every year. Every single day 900,000 cows, 3.8 million pigs, and 202 million chickens are killed for meat. Tracking the number of fish and crustaceans is far more difficult, but estimates put the number at about a trillion creatures killed each year. In the United States, just under 10 billion animals are killed each year, the vast majority of them the chickens that wind up in dishes like orange chicken. The average American carves about 220 pounds of meat every year from all those farmed animals, and the number goes up to 315 pounds if you include fish and other seafood, or 390 grams per day (about four Costco hot dogs).[5]

Producing a superabundance of meat and getting it to customers requires a complex value chain ranging from dedicating ever more land to growing feed crops through breeding, feeding, and slaughtering animals, and on to advertising and getting meat on shelves and menus. Getting here isn't simply a matter of farmers supplying consumer demand, but the result of a more than century-long process of developing the technology, policies, and business structures capable of mass-producing standardized meat and reshaping consumers' desires and tastes.

Tony Weis, a professor of geography at the University of Western Ontario in Canada, calls this transformation of the food system "meatification."[6]

Consider cows, arguably the least industrialized of the major meat animals eaten around the globe but also the most environmentally damaging. All cattle will at some point in their lives graze on grass, like the cows of Point Reyes. But the global cattle industry is a far cry from traditional livestock rearing. Although ranchers still raise cattle, many will send them for fattening at the sorts of grim feedlots photographed by Henner. Processing is controlled by a small number of large corporations, with four processors accounting for over 50 percent of all beef processing in the United States. This has been true since the centralization of slaughter and standardization of cuts in industrial processing facilities, immortalized in Upton Sinclair's *The Jungle*, in turn-of-the-twentieth-century Chicago. But today's slaughterhouses and processing facilities are located far from urban centers, out of the public's and regulators' sight, usually close to industrial feedlots in places like rural Texas, Kansas, Nebraska, and California; one set of twelve-wheelers brings in live cattle, and another set departs with refrigerated cuts of beef, while workers, mostly immigrants from Central America and Southeast Asia, turn the former into the latter. On feedlots, cows are fed a diet of processed feed, additives, and antibiotics, as are dairy cattle during most of their productive lives on large operations. That means that, to supply feed for meat and dairy, farmers must grow crops that are suitable for animal rather than human consumption.

Thirty-five percent of the country's corn becomes animal feed, as does most of its soy crop. Overall, 67 percent of all crops grown in the United States are used for animal feed (yes, you read that right), and only 27 percent of crop calories go directly to humans.[7]

All these animals and feed crops also guzzle scarce water. Animal agriculture's demand for water has grown so much that 70 percent of the water from the Colorado River is now used for animal feed like alfalfa. For comparison, the entire Phoenix metro area, home to five million people and now facing potential water shortages, gets about 40 percent of its water from the Colorado but accounts for only 5 percent of the river's water withdrawals.[8]

This large-scale remaking of what food is produced—where, by whom, and with what impacts—is what Weis means by meatification. Yet, as we mentioned, cows are the least industrialized of our most common livestock. Most of America's meat comes from pigs and chickens that will never feel sunlight and fresh air except on the truck that takes them to slaughter. These animals will live out their short lives on confined animal feeding operations (CAFOs, commonly known as factory farms). Chickens in particular have been genetically optimized to grow from chick to slaughter in mere weeks on a diet of processed feed in gigantic barns that can house tens of thousands of birds. If this sounds like the production of any other commodity, that's because it is. Chicken was once rare enough on American dinner plates that Herbert Hoover's promise of a chicken in every pot was a promise of wealth and abundance. Now Americans can eat chicken three meals a day. Public health scholar Ellen Silbergeld calls treating meat animals as if they were widgets the "chickenization" of the food system.[9] And consumers experience it in the ubiquity of chicken everywhere: on burgers, in salads, as wings and drumsticks, and as breaded nuggets and deep-fried blobs of orange chicken.

Meatification isn't just about the production of meat, but the production of consumers as well. The most obvious way it does this is by making a lot of meat very cheap. This meat is pushed onto menus and grocery aisles, what in the food business is called merchandising. But just to make sure, this is backed by massive advertising. Fast-food restaurants alone spend more than $5 billion per year on ads.[10] And the farmer-funded commodity-producer promotion groups known as checkoff programs also spend big money pushing pork (remember The Other White Meat campaign?) and milk (Got Milk?). In fact, dairy-industry groups have pushed the US Department of Agriculture to make milk a mandatory part of many subsidized school lunches.

But here's the depressing kicker. Like any capitalist enterprise, modern animal agriculture strives for efficiency, but while it may be an efficient way for large corporations to extract profit from producing a lot of meat, animal agriculture ultimately remains highly *inefficient* at producing calories and protein. Any animal will consume far more of these over its lifetime than its carcass will yield after slaughter. The conversion ratio of how many calories feed animals need to eat to how many of those calories end up on the plate as meat is a meager 13 percent for chicken (meaning that 87 percent of calories expended across the production process are metabolized by animals), and for beef it's a pathetic 3 percent. But what about protein, you ask? The numbers for protein conversion from feed to meat are just about as dismal: 21 percent for chicken and 3 percent for beef.[11]

That means that nearly all the land, water, and fuel that goes to making meat—the majority of the environmental impact of all of global agriculture!—doesn't wind up contributing to human nutrition. We would produce many more calories *and protein* for humans if we just grew crops to feed directly to people.

If we lived in a rational world, we would structure our agricultural system to be both as efficient and as *sustainable* as possible and use the incredible productivity of industrialized crop agriculture to feed people directly. Taken to its extreme limit, as it is by authors Troy Vettese and Drew Pendergrass in their left-wing utopian book *Half-Earth Socialism*, this idea would entail centrally planning agricultural production to maximize outputs and minimize inputs, leading to something like mandatory veganism.[12] We're not plant-based Stalinists ourselves, but it makes sense that even if we can't centrally plan agriculture, we would use policy to make it far more environmentally benign. And under capitalism, the best way of doing that is by making the worst parts of agriculture, like industrialized meat production, prohibitively expensive.

One idea popular in policy and economics circles for addressing harmful industries, be it oil and gas or tobacco, is to directly price in their so-called externalities, the harms they cause. The idea is that when the price of goods reflects their actual environmental and social costs, consumers are disincentivized from buying them. An attempt to model the so-called true cost of meat, published in 2022 in *Review of Environmental Economics and Policy*, suggested that the cost of meat should go up between 20 percent to 60 percent to account for its environmental impacts.[13] Similarly, in 2023 the German supermarket chain Penny, working with academic researchers, temporarily changed its in-store prices to reflect their estimated true costs. Sausage prices climbed 88 percent, cheese 74 percent, and dairy yogurt 34 percent. Predictably, sales of those items plummeted. The price of beans, broccoli, and tofu didn't budge.[14] If costs at the checkout reflected true costs, consumers would be incentivized to make more sustainable decisions, meaning far less meat.

Of course, people hate high food prices, and they hate taxes. That's because overt price interference feels like The Man rearranging their dinner plate, whereas government actions that drive prices down and shape what's for dinner, such as subsidies or weak regulations, are unseen and unquestioned. But even without new taxes, regulations that prevented the worst of the livestock industry's practices would have a similar effect. For instance, requiring factory farms to filter manure and use sewage systems rather than open-air lagoons would address a major source of pollution while forcing meat companies to foot the bill, which they'd pass on to the consumer. The same thing goes for requiring meat companies to treat animals better.

And framed this way, these regulations might be quite popular. A 2021 survey showed that 52.7 percent of Americans favored closing factory farms and that a whopping 49 percent were in favor of banning slaughterhouses.[15] Americans, in other words, despite being gluttons for meat, support quite radical anti-meat policies. And they will make their voices heard on the issue given the chance. Put to California's voters in 2018, Proposition 12 asked whether animal products sold in the state, no matter where they were produced, should meet a series of welfare standards. Prop 12 passed in a landslide over the organized objections of state- and national-level farm, retailer, and restaurant lobbies. And it survived court challenges all the way up to and including at the US Supreme Court.

But theoretical taxes and even Proposition 12—the most ambitious proanimal legislation ever passed in the United States—still only nibble at the edges of factory farming. Some activists want to push much further. On November 5, 2024, voters in Berkeley, California, opted overwhelmingly on a ballot initiative to ban factory farming in the city, making it the first jurisdiction in the country to make industrialized animal farming illegal. Of course, there are no factory farms in Berkeley, but just like the city's ban on the sale of fur in 2017, it shows the willingness of voters and activists to use the political process to shift norms around animal use. The Berkeley ballot initiative, and a similar one that was proposed but failed to pass in Sonoma, grew out of the work of Direct Action Everywhere (DxE), a Berkeley-based grassroots activist group. We first visited DxE a few years ago at its base of operations, on a leafy street south of the University of California, Berkeley campus, where a conference about plant-based proteins was underway. At the time, the DxE team was less concerned with dietary change and more with its nonviolent activism being targeted by the FBI.

DxE's relationship to the law can best be seen as an extension of the civil rights–era mantra of challenging unjust laws through nonviolent disobedience. In states like Utah that have passed "ag-gag" laws to prevent whistleblowing on farms, DxE blows the whistle anyway and dares the courts to hear its case. And if it finds sick or dying animals during its investigations, it removes them and tries to nurse them back to health. This is all technically trespass and theft, but DxE seeks its day in court to challenge the commodity status of animals under the law. While we were writing this book, DxE's efforts met with several surprising successes. Companies fled the court and prosecutors dropped cases at the prospect of having animal mistreatment admitted as evidence in court and exposed to the media. When DxE activists were criminally prosecuted in a ruby-red rural county in Utah, a jury unexpectedly cleared them of all charges.

But DxE's luck ran out in Sonoma County, just south of Point Reyes, in the beating heart of NorCal foodie culture. DxE often works to show that smaller-scale, putatively "humane" farming can have practically identical outcomes to industrial farming. In Sonoma, DxE rescued thirty-two ducks from Reichardt Duck Farm, an operation near Petaluma that, despite producing over a million ducks every year, markets itself to foodies as a family-run enterprise. DxE photographed dead, dying, and wounded ducks crammed into fetid barns—a far cry from the conditions described in glowing terms in *Sonoma Magazine* as ducks who "nest on comfy straw litter in an open environment, free to live their lives without interference."[16] At trial, a Democratic judge in deep-blue, progressive Sonoma County refused to allow any of that into evidence and also rejected, for reasons never made clear, an amicus brief from Harvard professor Kristen Stilt, one of the nation's leading experts on farmed-animal law.[17]

Wayne Hsiung, himself a lawyer and one of DxE's founders, was found guilty of criminal trespass. Much as Sonoma claims to be a paragon of small-scale farming, the reality is that its economy harbors extensive factory farming, which its voters, business interests, and judges are intent on protecting.

IF AGRIBUSINESS HAD ITS WAY, THE PUBLIC WOULD STOP ASKING SO many damn questions about how the sausage is made. But the story about livestock's agricultural impact is out. Despite the meat industry's best efforts at denying and downplaying its environmental and climate impacts, the public and policymakers are starting to take notice. Large farm operations and major processors are, not surprisingly, uninterested in being regulated or, for that matter, admitting responsibility for any environmental damage. But they are not necessarily averse to some mitigating measures if these won't hurt the bottom line and will generate positive publicity.

For example, dairy production produces a large amount of greenhouse gas (GHG) emissions, much of it from methane, generated both in cows' stomachs and from manure lagoons. The dairy industry has decided that just as you can sweep dirt under a rug, you can cover shit with a tarp. Literally. So-called biodigesters can be built over the lagoons, sucking up most (but not all) methane emissions and turning them into fuel for heating and vehicles. In major dairy-producing states like California and especially in major dairy-producing areas like the Central Valley, biodigesters have been subsidized by government green-energy schemes and have led to partnerships between megadairies and oil-and-gas giants like Shell and BP.

But biogas is not a miracle solution to the problems of animal farming. Instead, it illustrates how limited even the cleverest technical solutions are likely to be. For one, biodigesters address only lagoons, a tiny aspect of the full GHG account of dairy megafarms, where cows themselves still belch out methane and eat processed feed. Moreover, it's the larger operations that produce more methane, as well as more of other environmental harms, that benefit the most from biodigesters. As a result, biodigesters establish perverse incentivizes for farmers not to limit emissions but to grow the dirtiest elements of their operations because they make more manure, and therefore more cows, more profitable. Some estimates suggest that manure might now generate up to 40 percent of some dairy farms' revenue, leading critics to dub it "brown gold."[18]

And biodigesters are perhaps the least dubious of the attempts to mitigate Big Ag's climate hoofprint. Promises of seaweed-derived feed additives and even genetic engineering to reduce cows' emissions abound. These tend to generate headlines and hype around the possibility of climate-guilt-free meat. But despite big promises, there is simply no way to create a carbon-neutral cow, or anything close. Remember: cow, not how. Case in point: The Brazilian beef giant JBS, which has extensive operations in the USA and which has been linked to everything from deforestation to labor abuses to fraud in Brazil, claimed to be on track to be a net-zero company and—get this!—to being committed to "regenerative practices" despite as recently as 2022 having record emissions and *despite being the world's biggest beef company.* In

February 2024 the New York state attorney general sued the company for making misleading claims.[19] As of this writing, the court's decision is still pending.

If all of this sounds very close to the sorts of bait-and-switch tactics of delay and denial used by Big Tobacco and Big Oil, it's because that's where Big Meat got the idea.

SO IF A CLEAN COW IS AS MUCH OF A PIPE DREAM AS CLEAN COAL, going small-scale and free-range won't fix the problem, and regulation is an uphill battle, what should a concerned eater do?

The simplest answer is to eat less meat. Unsurprisingly, undoing meatification means demeatifying both our food system and our dinner plates to the extent possible.

A recent study published in the journal *Nature Food*—the imprint of the venerable science journal *Nature* dedicated to food science and policy—compared the CO_2 emissions and environmental footprint of different diets in the United Kingdom. It found that vegans had only a quarter the emissions and land use impact, and under half the impact on other environmental metrics, compared to those who ate a meat-heavy diet. Vegetarians also fared well, and the now-and-then meat eaters weren't as awful. But the resounding conclusion was that there exists "a strong relationship between the amount of animal-based foods in a diet and its environmental impact."[20]

The Lancet, the world's most prestigious medical journal, and EAT, a Swedish food system research NGO, reached a similar conclusion in 2019. Their collaborative project, the EAT-Lancet Commission, recommended a "planetary health diet" aimed at optimizing the nutrition and sustainability of diets.[21] It's capped at 2,500 calories per day and is rich in vegetables, nuts, pulses, fruit, and whole grains; is low in added sugar; and contains a modest amount of dairy, eggs, and meat. Something between a Mediterranean and a vegetarian diet.

The science is clear. But the politics of actually reducing meat consumption are complicated.

Depending on which statistics you look at, somewhere between 90 percent and 98 percent of Americans eat meat and aren't too keen on giving it up.[22] Of course, most people, including meat-obsessed foodie writers, will knowingly shake their heads at any mention of factory farms and acknowledge that yes, of course, those horrific things need to be done away with. But fewer people are willing to acknowledge the corollary of that conclusion: With the vast majority of meat that Americans consume coming from CAFOs, there would be 99 percent less chicken, 97 percent less pork, and, if we count industrial feedlots, upward of 65 percent less beef to go around. And that's not counting the milk from industrial dairies and eggs from industrial farms. If we closed all CAFOs overnight, most people would be vegan by default by tomorrow morning.

Less politically minded eaters would often just prefer to look the other way when it comes to the consequences of their appetites. Meanwhile, conservative politicians and activists have doused the issue in polarizing culture war accelerants to forestall even meager efforts to address global climate change through dietary change. These demagogues characterize the effort to reduce meat consumption as, variously, a globalist conspiracy to make people eat bugs and a Marxist plot to feminize American men. Faced with a polarized political landscape as toxic as a manure lagoon, even well-meaning commentators tiptoe away, equivocate, or try to change the subject ("Have you heard about how much water is involved in avocado production?").

Let's not tiptoe. Reducing your consumption of meat, and particularly beef, is the single most significant way you can reduce the environmental impact of your diet. Full stop.

But that observation must be the starting point of the conversation, and it can't be the end. Indeed, that observation alone sidesteps what kinds of support, individual and collective, make dietary change a gourmand's joyful feast rather than an ascetic's sour sacrifice.

And it can be joyful. This has long been the message of plant-forward cooks and those seeking to align how we eat with environmental goals. Frances Moore Lappé's 1971 book *Diet for a Small Planet* did exactly this: proposing delicious, often seasonal, vegetarian recipes to incentivize readers to expand their palates and their environmental consciousness. Mark Bittman, one of the foodie writers we've criticized in past chapters, has to his credit championed a "vegan before 6 p.m." diet backed by a cookbook designed to convince omnivores to explore diverse and nutritious plant-forward foods, at least until dinnertime. Peter Singer, the utilitarian philosopher who wrote the animal rights classic *Animal Liberation*, even included a recipe section in the book's first edition in 1975 to prove that giving up meat didn't mean giving up pleasure. Go to any bookstore today, and the range of cookbooks full of nutritious plant-based meals from around the world should at the very least tempt you to lean into meatless culinary diversity.

Indeed, for many, that may simply mean going back to parts of their culinary cultures as they were before they were meatified. One of Jan's most vivid childhood memories is helping his grandmother plant vegetables at the small country house outside Warsaw where he spent most of his early years. Small-scale gardening, gathering, and farming helped to supplement the provisions of Poland's state-managed food system. Potatoes, cabbage, carrots, peas, beets, and herbs like parsley and dill came from that garden. Then there were the fresh apples and pears and apricots. And wild berries and mushrooms to be had in the woods. From these would come entire dinners, bolstered by local products like flour or milk. *Pierogi ruskie* filled with potato and cheese, fresh *surówka*, pickles and sauerkraut, *kompot* to drink. Of course, there were eggs, and there was some meat, but not much—vastly less than what the current American or Polish diet provides. Eating any of these meals today doesn't feel like reinventing the wheel or making a culinary sacrifice. It feels like coming home.

You may now be saying that this is all well and good, but it comes down to navel gazing or, even worse, virtue signaling. How can vegan cookbooks address the environmental impact of food at scale? Individual changes, we are frequently told, are simply far too minimal to make progress against large systemic problems. This does contain both an important grain of truth—obviously forgoing a steak tonight isn't going to cause the meat industry to disappear by the morning—and a large helping of truthiness, or the gut feeling that something is right despite a lack of evidence, designed to assuage our consciences for our inaction on real problems.

But assuming a dichotomy between individual change and systemic change misses how the two are interconnected. We live in a society. And that means individual actions matter as part of changing norms, shifting market signals, and potentially building support for policy and political action. Humans are social creatures. We talk. We share ideas. And we share meals. And, of course, we then talk and share ideas about those meals. And in doing so, we shape social norms. Take veal. Even in a beef-loving country like the United States, veal is rarely eaten, and consumption is dwindling, in large part because of the social awareness of the cruelty in which veal calves are raised and, perhaps, squeamishness about eating a baby animal.

The removal of veal from most American menus reveals something that meat industry talking points often obscure: In a world where the effect of consumer choices is muddled by uncertainty and asymmetries of knowledge, altering your diet to reduce meat consumption may be the place where your individual agency as a consumer is most potent, particularly when it is reinforced by smart policies.

The meat industry's propaganda is uniquely designed to *obscure* the potency of dietary changes, as environmental policy scholars Lorendana Roy and Jennifer Jacquet argue.[23] Their research shows that meat companies and their apologists present current levels of meat consumption as an immutable law of nature beyond the influence of society, policy, and individual action, going after everything from national dietary guidelines to Meatless Mondays. That line about individual action not doing anything to address systemic problems in the food system that you've probably heard hundreds of times? In part, that's a meat industry talking point. The meat industry, as we mentioned earlier, borrows a lot of tactics from Big Oil, but consider this key difference between corporate messaging about fossil fuels and meat. Consumers are highly constrained when it comes to choosing how and why they consume energy in their homes, offices, and commutes, so oil and gas companies have long been pushing individual responsibility and carbon footprints to distract from this fact and from the need for major change to energy infrastructures. But when it comes to eating, where most consumers have considerable latitude for choice and agency multiple times every day—such as deciding to serve a three-bean chili instead of a chili con carne, or opt for the orange faux-chicken nuggets—meat companies have spent considerable resources trying to convince you that individual choice doesn't matter all that much. And that's even as they keep advertising meat to make sure you buy it. That the meat industry is invested in disempowering you as a consumer should tell you that they fear the consequences of consumers realizing they can (and should) choose otherwise.

Changes in consumer habits, even among disaggregated consumers living in places as far apart and as different as San Francisco, Indianapolis, and Baton Rouge, send market signals to retailers and producers. If enough people like the idea of replacing chicken nuggets with Beyond nuggets, retailers like Walmart and chains like Panda Express will be incentivized to match demand with supply and bring those products to California, Indiana, and Louisiana. And having those products in stores can, in turn, normalize them in the eyes of other consumers. The same goes for more demand for salads on menus and even salad-forward restaurants and restaurant chains.

But bigger changes in the impacts of consumption come not from disaggregated individual choice but from changing the options from which people choose their meals. Extensive research has shown that in institutional contexts where a large number of people eat large amounts of food, simply making more healthy or sustainable options the default can reduce environmental impact while improving consumers' diets. And even more research has shown that people often stick with default food options rather than making a fuss, so simple things like making vegetarian meals or even plant-based milks the norm, be it at weddings or conferences or dining halls, makes people's diets more sustainable without them even realizing it.

In San Francisco, this idea is being used in the school system, where a national initiative called the Good Food Purchasing Program is seeking to reshape how schools feed kids. The San Francisco Unified School District serves six million meals every year, and the Good Food Purchasing Program seeks to shift all the money spent on procurement for all those meals to buy and serve more local, sustainable, healthful, high-animal-welfare, and plant-forward options. It's a way to move away from chicken fingers, fries, and cartons of USDA-mandated milk to more healthful and varied meals. Although the shift to plant-based has not been large in any individual meal, or even always noticed, even small changes add up during six million meals per year. And we know that palates and food habits are shaped in youth, so such programs not only nudge students toward more sustainable and healthy eating but also prime them to be more conscientious eaters in the future.

All of this requires people to reduce meat, actively or passively, but it need not be a chore. And a new generation of entrepreneurs is cooking up ideas to make the shift to a more sustainable diet even more palatable.

PAT BROWN IS A NERD. AN EARNEST ONE. THE MOST ENDEARING SORT. FOR him, the food system is an equation. Each product requires certain inputs and has certain outputs. Given environmental constraints, it makes sense to not just produce food as efficiently as possible in general but also to focus on producing the most efficient kinds of food possible. All those statistics we threw at you earlier in this chapter? That's more or less how Pat understands the food system. He also happens to be a renowned scientist who came to food from an illustrious career in medicine and biochemistry. A longtime vegetarian for environmental reasons, he would look at agriculture and shake his head at the irrationality of it all.

"If you got rid of animals from the food system, you'd address most of its problems," Brown told us matter-of-factly. But Brown also knows that humans aren't always rational, especially when it comes to dinner. So how might we align irrational humans with a more rational ordering of the food system? The answer, for Brown, is to meet most people where they are, using the most efficient foods (plants) to give them the same flavors and eating experiences that they value from inefficient foods (meat). To explore that strategy, Brown left his professorship at Stanford and founded a company called Impossible Foods in 2011.

Impossible, which launched its first, eponymous burger in 2016, is an old hand by the standards of the so-called alternative-protein industry. Also known as "plant-based meat," products like Impossible's aim to re-create meat, or at least the taste, mouth feel, texture, and place on the plate of meat without using animals. The theory of change is admirably simple: to lower the switching costs for consumers to a more sustainable alternative. In economics-speak, "switching costs" refers to the expense in money, time, and effort in switching from one product to another. In the case of making more sustainable eating choices, switching costs can also take on profound gustatory dimensions.

Hence burgers made from soy. About 40 percent of all beef eaten in the United States is consumed in the form of ground beef, and 70 percent of all beef eaten in restaurants is as a burger. The notion of a "superfood" is nonsense made for advertising copy, but as far as super-nutritious foods go, you can't do much better than soy. The soybean is about 35 percent protein, and its protein is complete, containing nine amino acids that the human body cannot make on its own. And as an oily bean, it's easy to convert into other forms, like ground meal. It's also already widely produced and relatively cheap. For Brown it was a no-brainer to use soy as the primary building block for a better burger.

The trick was grinding and mixing it back together in a way that mimicked the cooking qualities and mouth feel of a beef burger. Methylcellulose, a binder used in jams and ice cream, and some basic starch held it all together, and sunflower and coconut oil give it fat for flavor and cooking. But it was missing a final touch: the flavor of meat! That particular metallic tinge comes from heme, the iron-packed molecular component. Impossible found that leghemoglobin, a protein found in the roots of soy, stimulated eaters' taste buds similarly to myoglobin, the protein in meat where heme is found. Impossible grows its leghemoglobin from a genetically engineered yeast—it's vastly cheaper than trying to extract it from soy roots—and adds it as a finishing dash to its burgers.

The result is a really good burger. We are fast-food purists. Bun, burger, a squirt of ketchup, and a dash of mustard. Maybe a pickle if we're feeling fancy, which we're usually not. Hold the white onions, you monsters. Cooked on a grill, still a bit smoky from the charcoal, an

Impossible burger is nothing short of meaty. It's everything you want from the most all-American of all-American dishes.

And that's not just our opinion. A 2019 taste test held for 175 participants at the Cornell University Sensory Evaluation Center, including both a blind and informed tasting, found that the majority of participants in *both* tests found the Impossible burger more delicious than a 100 percent beef burger. Committed beef eaters were fooled and convinced in equal measure.[24] That same year, conservative American media personality Glenn Beck did a live, blind taste test. After plucking the veggies off the burgers he was brought by his assistant—who says we can't agree on anything?—he savored both an Impossible and a beef burger and couldn't tell which was which. "That is amazing!" he exclaimed.

The burger eaten by Beck and the Cornell study participants and millions of consumers since emits upward of 90 percent less greenhouse gases than a beef burger, it uses less land and less water, and has less impact on biodiversity and soil quality. It is, in other words, a clear environmental win, but with all the flavor and pleasure of the original. "Common sense," according to Brown.

Virtually all alt-protein products use common plants as their principal ingredients. For all the big talk in the media—among both boosters and critics—of the technological novelty of products like the Impossible burger, ultimately they're beans, just prepared in such a way that they hit the same pleasure notes as meat. The disruption of the meat industry promised by the current crop of plant-based alt-protein companies is less rooted in Silicon Valley tech than it is in millennia of agricultural and culinary tradition.

Perhaps more than anywhere else in the world, this tradition can be traced back to China, which has a millennium-old documented history of vegetarian cooking. Buddhist and Daoist temples and adherents, including those that influenced the cuisine of the imperial courts, foregrounded tofu and bean curd in their cooking for centuries, turning soybeans into a wondrous diversity of dishes, both exquisite and afford-able, refined and rustic. Qing imperial chefs, inspired by these tradi-tions, emulated popular meat-based dishes, such as faux goose made from marinated tofu.

A trip to San Francisco's Chinatown will do a lot to rehabilitate tofu for those among the American public who see it as a tasteless, slimy goop choked down by pleasure-averse vegetarians. Back in the city from Point Reyes and in between visits to alt-protein startups, we stopped at a hole-in-the-wall joint where, on a weekday at lunch hour, every table was packed with eaters, many tucking into huge steaming bowls of noodle soup. We feasted happily on a selection of premodern alt-protein dishes. There was fried, perfectly crispy mock duck made from wheat gluten. There was fiercely spicy mapo tofu made with velvety silken tofu. There were almost translucent rice noodles drizzled with soy sauce and rice vin-egar and served with uncannily shrimpy mock shrimp made from konjac and seaweed. And then there was more tofu, mixed with mushrooms and folded into sheets of yet more pressed tofu.

The vegan food in Chinatown is a snapshot of democratic hedo-nism in action, at least in part. It's delectable, nutritious, affordable, made from mostly commonplace and sustainable ingredients, and clearly so popular that getting a table and some elbow room is an ask. It prompts the question of why we can't all just eat like this. And there's probably no single answer to that, ranging from availability to food cultures to different palates. But there is also much to say about taking this basic model and applying it to the American food context.

Again: hence burgers. Pat Brown understood that to make any meaningful environmental impact, his burgers would have to reach a wide range of consumers. And that meant not merely imitating meat but being the equivalent of meat across the metrics that matter most to consumers: taste, habit, and price. Although the Impossible burger launched to great fanfare at New York's much-lauded Momofuku in

2016, Brown's goal was never to create a curio for elite eaters. Rather, the idea was to use the industrial principles of economies-of-scale production to bring the burger to the masses. In a major coup, the company got its burgers on the national menu at Burger King in 2019, where they remain to this day.

If the United States has a truly homegrown national cuisine, it's fast food, an ingenious business model for turning mass-produced crops into standardized food designed to appeal to taste buds, wallets, and a need for convenience. Potatoes for fries, wheat for buns and breading, chicken for nuggets, and, the king of fast food, the burger. And it is the burger that has by far the largest environmental impact. On a research jaunt to the Central Valley, we saw dozens of feedlots and industrial megadairies producing the basic inputs for America's favorite mass-produced meals—each one an environmental disaster guzzling ever-more-scarce water and belching methane into the atmosphere. On the way back to San Francisco, we stopped for a couple Impossible Whoppers just off the I-5. Something so familiar and so completely different, salt and grease and all, but orders of magnitude more sustainable than a conventional Whopper. The Impossible burger, much like the Beyond orange chicken, slots right into all that is both amazing and awful about how Americans like to eat.

Although we shouldn't anchor our vision of progressive food on Burger King or even, for that matter, on burgers, in a narrow sense sustainability simply means doing less harm within existing systems: If every burger in US fast-food restaurants were replaced with an Impossible burger, it would be an environmental game changer. It's the sort of thing that can make you get philosophical in a strip mall parking lot on the outskirts of Lathrop, California, before finishing your watered-down Dr Pepper and driving to the Bay.

But even if one doesn't presume a fast-food model, the lesson of Impossible is that consumers are most likely to embrace plant-based alternatives to animal agriculture staples when those alternatives are configured to deliver the satisfactions they already know and crave. That's as true for meat alternatives as it is for the unsung heroes of the alternative-protein category: plant-based milks made from soy, almonds, oat, or more esoteric nuts and seeds like flax, macadamia, and pistachio. Starting from the same humble health food store beginnings as tofu, these alternatives, all of which have a much lower environmental footprint than dairy, have been surging. They now make up about 10 percent of US market share for liquid milk. Some coffee shops such as Blue Bottle have started using oat as the default. And as of the fall of 2024, Starbucks has stopped adding a "vegan tax" surcharge to plant-based milks. The switch to alternative milks may well be the lowest-hanging fruit in any eater's quest for sustainability.

Milk is easy, but cheese is hard. It is many people's stumbling block on the path to cutting out animal products, from the slice of American cheese that makes it a cheeseburger to feta in a Greek salad to unctuous camembert spread on a baguette to sharp blue cheese that serves as a test for the sophistication of one's palate to the red wax-wrapped Babybel tossed at the last minute into a kid's lunch box. Cheese has long been an alt-protein challenge, with the fermentation process seemingly too complex and the flavor notes too subtle to mimic with plants.

Overcoming this challenge was the task that Miyoko Schinner set for herself in the 1990s. A die-hard vegan—she has a tattoo on her toned right bicep that reads "phenomenally vegan"—she studied classical cheesemaking techniques and, through painstaking trial and error, began creating cheeses from nuts, seeds, and oils that were worthy of being paired with an artisanal baguette and a Sonoma wine. Miyoko's Creamery, the plant-based cheese company she founded, quickly became a cult favorite and then found itself on the shelves of first Whole Foods and then Trader Joe's.

We paid Miyoko a visit at Rancho Compasión, her gorgeous ranch in

Northern Marin County, less than an hour's drive south of Point Reyes. From the outside, the property looks like many of the small farms that dot the landscape in this part of California, except the animals at this vegan cheesemaker's ranch are not commodities. This ranch is an animal sanctuary, where animals, many rescued from commercial operations, are cared for and range freely. Miyoko is an entrepreneurial powerhouse and, although she visibly bristles at each of these terms, is the de facto godmother and ethical North Star of the US alt-protein industry. We sit on her couch playing with her dogs while she hauls in samples of some of the new cheeses she's whipping up in her kitchen. Miyoko insists that verisimilitude to dairy shouldn't be the point; she's honing her cheesemaking craft to create plant-based cheeses that are tasty on their own terms. Nibbling on her homemade samples, it's hard to disagree. They're *not* dairy-cheese analogs, but they're delectable, capturing a variety of complex flavors and memorable textures, from piquant and crumbly to smoky and velveteen, that would make any cheesemaker blush. All this *and* they are far more sustainable than the Point Reyes dairy cheeses made just up the coast.

But, paired with a tad bit of tech, plant-based cheese can indeed beat conventional cheese on its own terms. In early 2024 the Bay Area plant-based cheesemaker Climax Foods, which uses basic plant ingredients, expert cheesemaking, and a tiny dash of machine learning to mimic dairy cheese, was chosen as the winner of the Good Food Awards, a foodie prize, for its blue cheese. Climax never received the award. After some behind-the-scenes shenanigans—interference by dairy producers is widely suspected but not proven—the organizers retroactively changed the rules to disqualify Climax for using an ingredient not approved for retail sale in the United States. But Climax's near win showed that plant-based cheeses could deliver consumers authentically cheesy pleasures without the animals.[25]

* * *

ALTERNATIVE PROTEIN EXISTS BECAUSE DESPITE ANIMAL AGRICUL-
ture's system's awful outcomes for animals and the environment,
appeals to ethics and the environment alone fail to alter eaters' behav-
iors. For business as social change to succeed, agents of change need to
be entrepreneurs, selling consumers something they actually want to
eat. Few people in the industry capture that entrepreneurial gusto as
much as Josh Tetrick, the CEO of JUST.

If Pat Brown is the understated, calculated scientist who stays out of
the spotlight, Josh Tetrick is the opposite, the jock to Brown's nerd. A former football player with an athlete's build, square-jawed all-American good
looks, and a reputation as both a wooer of investors and a my-way-or-the-
highway boss, he is both media savvy and media fodder.

On the day we interview Tetrick at JUST's headquarters in Alameda, we run into him in the parking lot, and he is bleeding from a
cut sustained from mountain biking before work. He's still wearing his
biking kit, and he doesn't change for our meeting. But he does want
to talk. About his new bioreactors, his vision for his company, his 99
percent hold on the alternative-egg market, the food tech regulatory
environment in Singapore, and his partnership with world-famous chef
José Andrés. And he definitely wants us to try his cultivated chicken.
His ideas are big, his excitement contagious, and his track record as
bumpy as any other Silicon Valley golden boy. It's no wonder that an
entire book, Chase Purdy's *Billion Dollar Burger*, was written about the
so-called future of food with Tetrick as its de facto protagonist.[26]

Just to be clear: JUST doesn't make burgers. Rather, Tetrick
took the strategic approach of taking on animal-based products for
which there was ubiquitous demand and few if any analogs: mayonnaise and eggs. The United States produces more than ninety billion
eggs every year, and in one form or another they are everywhere and
in everything—omelets, breads, baked goods, pastas, batters, cake
mixes—and they're the main ingredient in America's favorite glop,

mayonnaise. Starting in 2011 with a company then called Hampton Creek, Tetrick launched an egg-free mayonnaise that quickly found its way into major stores like Target and saw venture capitalists lining up for a chance to get in on the ground floor.

But Hampton Creek grew too quickly. Some setbacks led to Target pulling the mayo and the board of directors abandoning Tetrick. He refocused and renamed his company JUST, a quaint double entendre for both justice and the company's new minimalist focus: producing a reliable substitute for eggs as an ingredient in home cooking. They found that substitute in the humble mung bean, its proteins isolated and, much as with the Impossible burger, mixed with flavoring and binders. The result is a thick yellow liquid sold in Tetrapak cartons that works just as well, as it were, in a breakfast burrito or a fall vegetable frittata with squash and russet potatoes. It's delicious and—and we hope you're starting to recognize this pattern—has a fraction of the environmental impact of conventional eggs. JUST now controls 99 percent of the plant-based alternative-egg market.

Still, JUST's success raised an interesting question: If you can have eggs without chicken, could you also have chicken . . . without the chicken? And we don't just mean an imitation. Could you produce real chicken flesh without the living bird?

A growing number of researchers, entrepreneurs, and investors are looking to technologies that allow tissue to be grown from live animal stem cells rather than entire animals. It's called cellular agriculture, and although it's been a staple of science fiction for ages, it was not even shown to be physically feasible until a 2005 academic paper. But the technology has grown by leaps and bounds since then.[27]

That's why Tetrick is so excited to show us his bioreactors: the giant steel vats, like beer tanks at a brewery, where the company's other major investment, he claims, will grow at scale. Under the subsidiary GOOD Meat, Tetrick's was the first company to get regulatory approval for a cellular-agriculture consumer product, launching its chicken in Singapore in 2020 and then in the United States in 2023. Though still priced as a high-end indulgence and sold at a loss, the GOOD chicken was a trailblazing achievement.

In Alameda, we try some: a plump, grilled morsel served on rice and accompanied by charred asparagus. It tastes, as the truism goes, just like chicken. Of course, it's not all chicken. It's a blend of plant ingredients and chicken cells, but we doubt you'd know it. Some sticklers claim they can tell a big difference; we sure as hell couldn't. Only a Cornell University Impossible-burger-style taste test will confirm what's real gustatory experience and what's bias.

While in San Francisco we visit a few other cellular agriculture startups less advanced than JUST. By far the most delicious is Wildtype, which is making cell-based salmon, a product with a different set of potential environmental upsides and a much higher price point than chicken. We try salmon and avocado rolls prepared by a sushi chef at the company's loft office overlooking the San Francisco Bay. It's nothing short of astonishing. The salmon looks the part, with striations of fat lining pink flesh. And although it is also made with a mix of animal cells and plant-based ingredients, it tastes just like salmon. Good salmon.

If eating a bowl of Beyond orange chicken or an Impossible burger can excite your imagination about tasty ways to reduce the environmental impact of cherished dishes, cellular agriculture poses even more profound philosophical questions—not by offering the same dishes from different ingredients, but by dishing out identical ingredients using radically different production processes. In 2008, when the technology was still largely speculative, philosophers Patrick Hopkins and Austin Dacey argued that cellular agriculture was a moral good specifically because it sidestepped debates about the desirability or palatability of meat; it changed physical reality rather than people's minds. Appealing to people's "selfish" pleasures was, they argued, both eminently practical and ethically sound.[28]

We'd argue that by decoupling the pleasure of meat from the harm it causes animals and the environment, cellular agriculture is literally democratic hedonism given flesh.

In theory, cellular agriculture gives people meat, just produced through different methods that burden society, animals, and the environment far less. It reduces the costs of switching from conventional meat to a planet-friendly alternative right down to zero on every tangible metric, leaving only intangibles, like unease with technology and a longing for naturalness. Yet as we've described, the process of meatification was not natural—modern abattoirs are as distant from nomadic herding and hunting as your cell phone is from an ax head. In fact, cellular agriculture requires purity of cells and production processes that are materially vastly purer than conventional meat—think no *E. coli* in chicken from feces in a slaughterhouse and no heavy metal residues in fish. Wildtype's CEO tells us that the spark that lit his desire for cell-ag salmon wasn't an ethical concern for fish—a lifelong fisherman, he does hope his product lets wild fish populations recover from overfishing—but the wish that his then-pregnant wife could eat sushi without worrying about mercury contamination.

A few years later, that dream is closer to fruition. In 2025 Wildtype became the first cultivated fish company to gain FDA approval for commercial sale, launching at Kann, an upscale, award-winning Haitian restaurant in Portland, Oregon. We had to be there. The dish was a riff on ceviche, with small cubes of cold-smoked Wildtype salmon mixed with diced pickled strawberry served with a fried epis rice cracker. It was perfectly salmony in flavor if not entirely in texture. And it was pricey, but not as pricey as many of the conventional meat dishes on the menu. It's a big step: a cultivated fish deemed perfectly safe by regulators being scarfed down by curious eaters.

Comparing the environmental impacts of an emerging product to existing conventional meat is tricky. By not using animals, cellular agriculture requires less land and almost certainly less water. But the energy required to run bioreactors is considerable, meaning that the technology would be a decisive climate win only if it were powered with renewable sources of energy, and even then it would depend on which meat it was displacing. (Chicken is substantially less GHG-intensive than beef.) The bigger and more immediate problem, however, is that, prototypes and high-end tasting dinners aside, cellular agriculture is not yet ready for the mass market, and it's unclear when, if ever, it will be. Technological roadblocks abound: from bioreactor design to less expensive growth mediums to processes to create appealing cuts of meat. The technology may well have been commercialized prematurely. Without access to adequate support in academia and government labs, scientists have had to rely on venture capital to fund basic research, a dynamic that has driven hasty paths to commercialization and hyped prototypes, with no products consistently available to everyday consumers. Only in recent years is this gap being addressed, with major university research centers, funded by the USDA and other partners, being set up at the University of California, Davis, and Tufts in Boston. This means that the technology may well come to fruition, but it's a matter of decades and not years for viable mass-market products to find their way onto our plates.

ON JANUARY 8, 2025, POINT REYES CHANGED FOREVER. AFTER YEARS of negotiations involving local stakeholders and major NGOs like the Center for Biological Diversity and Nature Conservancy, eleven of the area's dairies will allow themselves to be bought out, giving up their leases on the land. Some cattle ranches will remain, but 16,000 acres of land will be allowed to rewild, opening access to those native tule elk.

It's a decision which tacitly acknowledges that the sorts of farms whose boosters claimed worked in harmony with nature were in fact inimical to it.

Agriculture is a huge source of environmental harms, and they are harms in which we, as eaters, are all complicit. The scale of the problem and the necessary scale of the solutions can be daunting. That may, in part, explain the pull of the ideal of small, local farms and of a return to the land. But to make a dent in the food system, you need to change how people eat at a huge scale. The scientific consensus points us squarely in one and only one direction: eating less animal-derived products and more plants, be they fresh, cooked, or turned into a milk, meat, cheese, or egg alternative.

Yet the food system is far from rational. As we wrote this book, the market for plant-based meat, which had been surging, retracted. Companies with products at major supermarkets went bankrupt or shrank their product offerings. Inflation hasn't helped. Neither has a growing tide of pushback against alt protein, originating both from the incumbent meat industry and from the food press. Some of this is organic (pun intended) concern with the naturalness and health benefits of the foods. But much is bad-faith critique, some of it fed by organized efforts by the meat industry and its front groups. Meanwhile, lawmakers aligned with meat interests have pushed for legislation that would prevent alt protein from using terms like "meat" to describe their product, and, more drastically, states including Florida and Alabama have banned cellular agriculture altogether.

If you build it, will they eat it? Is it just price, taste, and habit? The future of alternative protein is unwritten. There is massive demand for more sustainable and delicious food, and alt protein is now a major market category where, ten years ago, there was none. But simply relying on market mechanisms or politics as usual will not get us there, especially given the political and social pushback over something as sensitive as food and as cherished as meat. Ultimately, the potential transformative synergies among activists, consumers, and policymakers are as much art as science. This is a chicken-and-egg problem. Can policies positively alter consumer behavior, or do consumer preferences drive policy and political outcomes? The answer is . . . *yes*. Both are needed, as are boundary-testing activists like DxE that are willing to say unpopular but important things and unsettle what consumers take for granted about the food system. But for all these challenges, changing diets remains the singularly most powerful way to reduce the environmental consequences of food production.

Chapter 4

LUNCH LADY POLITICS

NEW YORK CITY IS UNDENIABLY ONE OF THE GREAT GLOBAL FOOD cities. The greatest, if you ask most New Yorkers. Every neighborhood across the five boroughs—and in some neighborhoods every block—has its own culinary traditions and go-tos. From dim sum in Flushing to Ital juice-and-patty joints in Crown Heights to pierogi in Greenpoint to New York–style soul food in Harlem, the city exudes food culture. It is the hometown of countless globally renowned restaurants, from the homey deli Zabar's to trendsetting Momofuku to the institution that is Eleven Madison Park, and on to countless others etched in culinary history. NYC's seventy-one Michelin-starred restaurants rub shoulders with the ubiquitous bodegas with their bacon-egg-and-cheese rolls. The city has also birthed generations of guides to its foodways, including perhaps the most famous American food raconteur of all: the straight-talking, drug-addled, jiu-jitsu-fighting, tattooed culinary Herodotus, Anthony Bourdain.

The city's food culture, like the city itself, is—it's a cliché, but it's true—a melting pot. It's diverse, cosmopolitan, pluralistic, a place where these American ideals are so organically lived and embraced that it seems almost a place apart or, as the playwright and die-hard New York nationalist Spalding Gray so pithily put it, "an island off the coast of the United States."

But in other ways the Big Apple is exactly like much of America: It is blighted by economic inequality and expensive rent and food. In one of the richest cities in the country, 14 percent of the population regularly experiences food insecurity.[1] For them, eating in NYC is less about scoring a hard-to-get reservation than it is about getting enough food to make it through the month.

For a country as wealthy as the United States, food insecurity is a surprisingly common problem. As many as 44 million people lack reliable access to adequately nutritious diets every year. Food insecurity does not just mean hunger but also a slate of issues, including stress, anxiety, depression, and diet-related diseases caused by eating, out of necessity, the wrong kinds of food in the wrong proportions at the wrong times.

A vital part of food security is food access, which means people being able to regularly eat a nutritious diet; in even simpler terms, it means getting the food that is available. The biggest barrier to that is money, which means the most direct solution to food insecurity is either giving people food directly or freeing up household budgets so that poor and working people have more money to spend on food. It's a simple and time-tested solution.

Foodie culture and conventional food writing tends to valorize visionaries, gurus, and mavericks throwing off the shackles of the food system. Food rebels are fun to read about—we've written about some of them in this very book!—but flashy ideas aren't what will keep millions of hungry people reliably fed. In fact, improving and expanding the existing programs aimed at reducing poverty and food insecurity is the surest route to making sure that everyone can have a delicious and healthy diet. In this chapter we focus on two programs—SNAP and universal school lunches—that have done much to alleviate food insecurity and improve access, even as they both could be improved in important ways. We also look at two communities of different sizes—New York City and Durham, North Carolina, both foodie meccas plagued by food insecurity—to examine the drivers of food insecurity and the sorts of policies that can (and can't) help people in need.

We converge on one fact: If people are hungry, the best solution is to find ways to feed them, no questions asked. And doing so at the scale of the problem requires large-scale organization, large budgets, solid peer-reviewed research, and tapping into the economies-of-scale food availability and low costs afforded by the conventional food value chain. We call it lunch lady politics.

IN THE MIDDLE OF THE SOULLESS URBAN NOTHINGSCAPE THAT IS MID-town Manhattan sits Le Bernardin, a French seafood restaurant awarded three Michelin stars, and long considered one of the finest in the city, if not the whole country. A chef's tasting menu with wine pairing will run you just about $500 a person before tip, and the spot has long been a haunt for wealthy Manhattanites as well as a must-visit destination for traveling foodies. Its chef, Eric Ripert, once gave an interview to Yahoo Finance that the outlet ostentatiously—but also correctly—headlined "How to Eat like the 1%."[2] Dinner at Le Bernardin is a pleasure well suited to New York City, a place where every twenty-fifth person is a millionaire and where sixty billionaires make their homes.

But this great wealth coexists with extensive poverty. For all the glitz and glamour, the median household income in New York City is about $70,000, just below the national average, while costs of living are soaring. The single biggest budget item for most households is rent, mortgage payments, or property taxes, and property prices and rents in New York City have done nothing but climb, undeterred by COVID, crime rates, or even renter occupancy. Manhattan, the richest borough, has an $85,000 median household income, but average rent for a one-bedroom apartment is $4,400. In Brooklyn, where Jan works, those numbers are $70,000 and $2,800. In the Bronx, $43,000 and $2,400. Remember the so-called 30 percent rule of budgeting a third of your after-tax income for rent? Or the "50/30/20 rule," where the 50 means budgeting half of your income for "needs" like rent and food? New York City's real estate market takes these rules, folds them into paper airplanes, and throws them into the Hudson. According to a United Way report that estimates the "true cost" of living in the city, 50 percent of all households across NYC don't earn enough to regularly cover all their monthly expenses.[3]

The USDA releases periodic nutritional guidelines for Americans that sketch out a model diet that would meet most people's nutritional needs. It also maps out budgets, based on national food cost averages, for meeting these targets with different foods at four price points: thrifty, low-cost, moderate, and liberal. Thrifty is ostensibly the cheapest possible way to ensure nutritionally sound food access. In late 2024 the USDA estimated the average monthly food cost under the thrifty plan for a family of four at $980 (or two tasting menus at Le Bernardin, for those keeping score). For a woman over fifty, it suggested $226. But in New York City, that doesn't come close to cutting it.[4] We priced out a month's worth of thrifty-plan-compliant food using the exact products the agency uses to make its estimates—including everything from frozen spinach through canned beans and on to staples like rice and pasta as well as some beef, chicken, eggs, condiments, and coffee—at a few of the city's more affordable supermarkets across the five boroughs. The average price came out to $1,150 for a family of four and $283 for a woman over fifty. If the poorest families and individuals in the city eat nothing outside the house and follow the tightest budget, they still overshoot the USDA minimum budget by 17 percent and 25 percent, respectively.

For many New Yorkers, the math just doesn't work.

We meet Maria Montes (not her real name), a retired legal secretary originally from the Dominican Republic, at the Holyrood Episcopal Church in Washington Heights. It's a sunny summer Tuesday, and Maria, a volunteer, is unloading boxes of food for the West Side Campaign Against Hunger (WSCAH), a food bank. The food will be distributed to the dozens of people patiently waiting in the church's pews. In exchange for volunteering, she gets to take her allotment first: a carton of milk, a bag of nonperishable foods, and a bag of fruits and veggies. Despite her pension and Supplemental Nutrition Assistance Program benefits, sometimes she runs out of food money, especially in the high heat of the summer, when her electricity bill tops $200 a month from running the air conditioner. And as her rent creeps up, Maria is noticing that gentrification is finally arriving in the northern tip of Manhattan she calls home. "Everything is more expensive every day," she says. It's part of the reason there are more than a million-and-a-half food-insecure New Yorkers like Maria.

SNAP is the nation's first line of defense against food insecurity. The program has its roots in the Great Depression, when economic catastrophe drove unemployment to almost 23 percent, sending countless millions of Americans into penury and hunger. The Depression prompted an unprecedented government response: Franklin Roosevelt's New Deal. In the case of food, the Federal Surplus Commodities Corporation was set up to be, as its head administrator Milo Perkins described it, "a bridge across that chasm" between farmers whose products were rotting in silos because of cratering demand and consumers who couldn't afford to eat. It bought, at wholesale prices, food from farmers that they couldn't sell in the open market and distributed it to the unemployed and to schools, thereby helping farmers and keeping the needy fed. This program evolved into the food stamp program, whereby the millions of poor people who got government financial support could buy orange stamps they could redeem for food and, for every dollar spent, would get an extra 50 cents' worth of blue stamps to spend on farmers' surplus, meaning mostly staples like eggs and beans and perishable foods like vegetables. The program increased their purchasing power, ensured that they spent money on food by converting cash into stamps, and kept farmers in business.

The program continued in different forms until after World War II, but then was shuttered during the postwar economic boom, only to be relaunched in the 1960s as a part of Lyndon Johnson's "war on poverty," first as a pilot program and then as a permanent national program in 1964. No longer designed to create a market for surplus production, it now linked the hungry with retailers, who would benefit from government-stoked demand. The program is now administered by the USDA and is bargained over every time a farm bill is passed, often with funding for the poor predicated on subsidies for farm interests.

After Medicare and Medicaid, the food stamp program, renamed SNAP in 2008, is the biggest federal investment in public health and poverty reduction. At any given moment, more than 40 million people in the United States receive SNAP benefits, at a cost of about $120 billion per year. Monthly SNAP payments range from $127 in places with lower food and living costs like New Mexico to $212 in New York City. It may not seem like much, but it can mean that recipients are able to eat three meals a day, as opposed to only two.[5]

According to the latest figures and peer-reviewed studies, SNAP often lifts as many as three million people per year—including one million children—out of poverty. Children in SNAP households are less likely to be food insecure, obese, or underweight. They are also less likely to be hospitalized or fall ill, which reduces health-care costs for their families.

SNAP, of course, has its critics. The conservative criticism is that food stamps are a bad idea in and of themselves because recipients should bear personal responsibility for their own hunger. This is both a morally bankrupt position and one that misses the basic economics of SNAP: It takes up less than 2 percent of government spending, a relative pittance for the broad benefits it provides, but it also goes to people who will spend all of it at retailers. It's a de facto subsidy for the retail sector and the food industry in addition to being a lifeline for the hungry. It's good politics, good policy, and good for business.

Too good for some, in fact. One criticism of SNAP we take seriously is that it doesn't actively encourage healthy food choices. When food stamps were formalized in 1964, some argued that not all foods should be made available, and specifically that soda drinks should be excluded. At the time, government officials were far more concerned about malnutrition than obesity, and so no restrictions were put in place. Sixty years after the program's launch, obesity is a far greater problem in the United States, and SNAP beneficiaries' single biggest expenditure is soda, with sugar-sweetened beverages accounting for about 9 percent of all SNAP spending.[6]

In 2011 New York City's mayor and anti-sugar activist Michael Bloomberg tried to temporarily prevent SNAP recipients from being able to buy soda in the city, ostensibly as a pilot project to measure the effects on population-level obesity rates. The USDA refused. It has refused similar petitions from California and Maine. The reason for the USDA's refusal is an interesting example of the odd bedfellows created by food policy and the perplexing array of constituencies to which the USDA answers.[7] Soda companies are, of course, the biggest beneficiaries of soda's availability through SNAP, just as other food companies generate enormous income from the program. For example, Kraft Foods, maker of the famous mac and cheese, sells a lot of food to SNAP recipients, and although it's hard to pinpoint a precise number, especially recently, a 2012 article suggested the number might be over 15 percent of all of Kraft's US revenue. Ironically, soda and snack companies have found unlikely allies in anti-hunger advocates and other progressives who believe that food access assistance should allow food-insecure people to make their own choices, unburdened by stigma or restrictions.[8]

In theory, spending 9 percent of SNAP monies on something that is as caustic for human health as sugary soda is very ill-conceived. But this may be one of those cases where the bad has to be taken with the good if the program is to work efficiently and with broad political support, even if Big Food wins as well. Beyond this, soda drinks and other sugary foods can have a place in a healthy diet if consumed in moderation, and if they were excluded from SNAP, beneficiaries would likely buy them anyway, clawing the money from their household budgets and defeating the purpose of the program. Denying people pleasurable foods under the guise of paternalism would therefore likely be self-defeating.

In any case, the most salient criticism of the program is not that it's too generous with its benefits; it's that for many working people, SNAP simply isn't enough and sometimes isn't even an option. In cities with high costs of living, like New York, people with full-time jobs may barely miss out on SNAP eligibility despite being on the verge of food insecurity.

To see what we mean, follow the math: In New York City, minimum wage is $16 per hour. Assuming a 35-hour workweek, that amounts to about $2,250 a month, or $27,000 per year. For a family of four with two minimum-wage-earning parents, that's $54,000 a year before taxes, or about $15,000 over SNAP eligibility. Assuming that this family spends the 13 percent of its income that the Bureau of Labor Statistics says the average New York household spends on food, that's $7,000 on food for the year, or $585 per month.[9] For those keeping score, that's about the same as one tasting menu including wine and tip for one person at Le Bernardin. Stop for a second and budget out 30 days' worth of meals for your family on under $600 a month. It's impossible, so families spend more on food and find ways to spend less on other things. And although SNAP benefits change depending on cost of living, SNAP eligibility doesn't. Changing this by making eligibility a sliding scale relative to cost of living would allow the program to help tens of millions more Americans.

Meanwhile, a family of four in NYC with a gross annual income of $39,000 or less can get up to $973 a month in SNAP. That's almost $12,000 a year in food assistance, enough to technically pull the family above the poverty line. They may still not have enough disposable income to spend on much other than rent and necessities, but they have the SNAP lifeline. If they supplement SNAP with other money—let's use that 13 percent of income here as well, so about $425 a month—that's $1,400 a month. That's enough to afford what we priced out the USDA's thrifty basket of groceries to be, with a bit of margin left over. It's not bad, but it's far from ideal. Treat this as real life and not just math: Budget in the odd splurge for coffee with friends or takeout when you're too tired to cook. Or even plan out a month's worth of breakfasts, lunches, and dinners varied enough that you can afford *and* that the kids will actually eat. And then there are unexpected expenses. Often, even the best-planned budgets are off, and the last few days of the month can be a stretch.

Even with SNAP, there are many reasons that people might experience sudden or exacerbated food insecurity. Much of it has to do with disrupted budgets. It might just be that SNAP or savings run out a few days short. Or there might be the sudden loss of a job. There's inflation. The best-made food budgets can be upset when average food prices jump almost 4 percent in one month, as they did in September 2023. Maria Montes, the retired legal secretary, uses plantains, a staple in her kitchen, as her example: "You used to be able to get, like, twenty for like a dollar. Now it's five for three dollars." Her strategies for making her SNAP money go further include asking her son to drive her across the George Washington Bridge to New Jersey and down to the Walmart Supercenter in North Bergen, where prices are vastly lower than in the city. Her son, who happens to work in the food industry in the city, has also covered her bills when she's come up short. She's glad for the help, but she speaks about it with wounded pride.

Food banks are a lifeline for people in crisis moments. There are five hundred food pantries and one hundred soup kitchens in New York City. The West Side Campaign Against Hunger is among the largest and oldest. Founded in 1979, it now serves more than twenty thousand people every year across the city, with a focus on fresh and healthy food. Once a month, customers—the pantry makes a point of referring to people whom they serve as "customers" and not aid recipients—get about three days' worth of food. It adds up to just over five million pounds of food distributed every year.

It's a gargantuan effort to get all that food to the people who need it. We spent a morning with the WSCAH crew, starting at their warehouse in a converted post office on 180th Street. With its large loading dock, huge industrial freezer and fridge, and rows upon rows of pallets of bags of oats, cans of beans, boxes of pasta, and other staples, it looks like a food service company's warehouse. Which, in effect, it is. Only it's funded by philanthropy, and, beyond a core staff, the work of sorting, packing, and distributing food is done by volunteers. At 180th Street, warehouse manager Steven runs a seamless logistical operation that is more modern fulfillment center than food bank, with orders color coded by diet and family size and an industrial-style assembly line in place to assemble food packages to be shipped out. What makes it all feasible is using donated money to buy strategically from wholesale distributors. It's not local or fancy food, and it's definitely not those feel-good individual food donations of a few cans of beans that make it work; it's large financial donations that allow bulk buying.

Studies of food-pantry use, including some that specifically reference WSCAH's work, show that it contributes to increasing fruit and vegetable consumption among food-insecure people and especially to improving child nutrition in food-insecure families.[10]

But it's still just a stopgap measure. Alyson Rosenthal, the chief program officer at WSCAH, has no illusions about the program's limits. "It's simply just not enough. A lot of people go to multiple pantries every month. For a lot of New Yorkers, it just becomes part of the way they make a go of life in the city," she told us. The organization also grapples with the question of whether to feed fewer people more often or to spread resources among more people fed less often. Rising food costs are also a problem, even at wholesale prices. "It comes down to budget."

Now, poke around New York City food politics long enough, and another solution to food insecurity will inevitably be brought up: urban gardens. Walk the streets of the city's boroughs, and sooner or later you'll run into a part of a city block not filled with concrete. Be it between two brownstones in Brooklyn or in the shadow of newly built condos in the Bronx, you'll see a patch of green, often filled with planters brimming with vegetables, sometimes marked with hand-lettered wood signs. They are beautiful, bespoke incongruities in the gray cityscape. New York City is home to hundreds of community gardens, at least 550 of which are supported by the city's GreenThumb program.

The gardens have a storied past of reclaiming urban space for communities and being sites of resistance to unchecked development, including pitched battles between gardeners and city hall over land leases. Studies of urban gardens consistently show their myriad benefits, ranging from the health perks of physical work outside to creating stronger communities and acting as spaces to incubate and act on the principles of direct democracy. And sometimes it's just good to get some dirt under your fingernails. It's beautiful to harvest something. Community gardens are an invaluable part of life in the city.

But one thing community gardens are not great at is improving food access, no matter how often this claim gets repeated.

Sure, producing your own food, especially in neighborhoods where ready access to fresh food is lacking, is a good thing. However, the problem in the city, and much of the country, isn't that there is an insufficient supply of tomatoes and cucumbers (availability) but that the people who should be eating the tomatoes and cucumbers cannot afford to buy them (access). Free or cheap locally produced food seems to be just what's needed, and to the extent community gardens provide produce to needy populations, that's fantastic. But it's just not a transformative intervention at any meaningful scale. Statistics for production quantities in these small gardens are hard to come by, but a 2012 report of 106 participating gardens showed a total annual yield of 87,000 pounds of food.[11] For the sake of argument, let's multiply that by 5 to match the official number of the city's gardens and assume they produce 435,000 pounds of food every year. An impressive bounty to be sure, especially when grown by hobbyists on small plots, but less than 10 percent of all the food distributed by WSCAH alone. And while it means that some people who might not be able to afford those tomatoes or cucumbers do get some now and then, New York City residents consume about 20 billion pounds of food every year. This means that even with our generous estimates, community gardens produce .002 percent (2 one-thousandths of one percentage point) of the city's food supply. Qualitatively invaluable, quantitatively valueless. Harsh but true.

We will defend community gardens to the death as crucial sites for community building, gardening education, and respite from the crush of the city. But if we're being serious about addressing food access, we must look for pathways like SNAP and food pantries that connect the affordable food grown on large-scale, productive farms to the food-insecure consumers who need it. But the most important place where that principle can be put into action is in helping the most helpless victims of food insecurity: kids.

IT'S A FROSTY FRIDAY IN EARLY DECEMBER WHEN WE VISIT PS56 IN Norwood Heights in the Bronx, one of the city's poorest neighborhoods and among those with the highest percentage of first-generation immigrants. But inside the school's turquoise lunchroom, the midday meal is a joyous affair. The kids, kindergarteners through fifth graders, come in in shifts and line up for their lunches, which are served by lunch ladies, including Rosa, a cheerful woman with a heavy Spanish accent who has been at the school for six years. It's Friday, which means that Rosa fills the center of the kids' plates with their choice of a grilled-cheese sandwich or a bean burrito, and then heaps on the sides: pinto beans slow cooked with veggies, roasted cauliflower, rice, and sliced apple. There's also a salad bar that, to our surprise, is mobbed. The students, seated at round wood tables, scarf down their lunches, chatting and laughing, all under the watchful eye of teachers who make sure the sliced apples don't turn into projectiles, which they do only once that Friday. As one group of students vacates the lunchroom, the kitchen staff move with incredible speed to clean up, prep a new stack of sandwiches, refill the trays of sides, and get ready for the next wave. In its industrial kitchen the school employs six cooks as well as the school's food-support staff, who have been there since six or seven in the morning, preparing and distributing breakfasts and cooking enough food to feed lunch to six hundred children. Paid for by the Department of Agriculture, the breakfasts and lunches are free, taking pressure off parents' wallets and schedules while keeping the students well-fed. At PS56, amid the laughter and chatter of the lunchroom, it's easy to forget that you're looking at probably the single most effective food-insecurity solution ever invented.

Decades of studies show that kids who go to school hungry or come from food-insecure homes do worse at school. They have trouble concentrating, develop writing and reading skills more slowly, and are more prone to behavioral problems, including anxiety and aggression. They also tend to underperform compared to their food-secure peers, which exacerbates all these problems, especially if children are treated as inferior from a young age by schoolmates, teachers, and administrators through no fault of their own. All of this, in turn, is both exacerbated by *and* exacerbates stress at home and potential social stigma caused by food insecurity. And children who suffer from all these problems are more likely to underachieve economically as adults, increasing their risk for long-term food insecurity. It's a vicious cycle, and it's one that should be completely avoidable in rich countries like ours. A team of researchers who ran one of the biggest metastudies on the relationship between food insecurity and child development concluded, in uncommonly blunt terms for an academic paper, that the fact that "children in countries producing a surfeit of food are denied the right to quality food is untenable and indicates a failure of political and public will."[12]

Giving kids food is the most obvious solution to the educational harms of hunger, especially since they are physically present five days a week in an institution that has the capacity to feed them: schools. The free school meal is the most effective policy for addressing child food insecurity because it addresses both the problem and the symptoms—though, of course, not the causes—of food insecurity, and it does it at the level of an entire affected population.

With a huge population of hungry kids, New York City has long been a laboratory for school lunch programs. At the turn of the twentieth century, it was a rapidly growing metropolis of three and a half million people, the largest commercial hub on the East Coast, and the first port of call for immigrants arriving at Ellis Island. That made it dynamic and cosmopolitan, but many New Yorkers lived in grinding poverty. Child malnutrition was rampant. In the absence of any government social safety net, it was up to families, including many large, newly arrived, and often not-yet-English-speaking families, to find ways to keep their children fed. Many couldn't.

Humanitarians and social reformers like Mabel Hyde Kitteridge organized to address this problem as best they could. In 1908 Kitteridge, the Boston-born daughter of an affluent pastor, founded the School Lunch Committee (SLC), a philanthropically funded organization based in Hell's Kitchen that provided soup and bread to local students free of charge. To get kids to eat, Kitteridge aimed for culturally recognizable meals for the Irish, Italian, and other immigrant groups, even hiring an Italian chef for Little Italy's public schools. By the end of World War I, as New York's school population ballooned, the SLC was serving eighty thousand meals per year across eight public schools. But as the program expanded, it grew insolvent and was ceded to the city's Board of Education, which introduced payments for meals. Participation plummeted.[13]

In the early twentieth century, a few cities, such as Boston and Philadelphia, implemented free school lunches, but it wasn't until 1946 that the federal government passed the National School Lunch Program under the auspices of the USDA. The NSLP, like early food stamp initiatives, was initially a way for farmers to sell excess products, especially milk and cheese, to school boards. It wasn't until 1962 that the NSLP became what it is today: a reimbursement program permanently funded through the national budget for the most needy children. As is often the case, the United States was stumbling, despite itself, closer to a commonsense welfare state policy.

Once you decide on funding a school lunch or breakfast program, however, the question becomes this: Who gets to eat? And what do they get to eat? Because, as with SNAP, federal standards for qualification and funds per student don't change from state to state or city to city, in some cities, like New York, many families who did not qualify for free or discounted school lunches or breakfasts were nonetheless food insecure. Meanwhile, food costs in major metropolitan areas tend to be higher, meaning that the government's per-student payouts don't stretch as far, resulting in worse meals or school districts having to foot the difference from their own budgets. As with many food-access issues, this is a problem that could be fixed with more money, but the rise of free school lunches in the United States collided with the economic stagnation of the 1970s and then the austerity measures of Ronald Reagan. Yes, the NSLP itself is permanently funded, but even modest budget cuts to local school districts can leave school kitchens understaffed or without working fridges or enough stoves.

Food access is a budgeting issue, no matter who does the budgeting or foots the bill.

Budget shortfalls generated two unfortunate consequences. First, food got worse. Second, food companies leaped into the budget breach with lower-cost, branded junk food products like snacks and soda. Caught between these two, schools also faced the problem of serving meals kids wanted to eat. Emulating fast food, by then America's national cuisine, was a makeshift solution to palatability and budget austerity that had awful results. Food writer Mimi Sheraton described the school lunches in New York City public schools in 1976 as made up of "tough, sodden hamburgers that are pasty and usually bitingly salty; limply breaded fried chicken; and salty hot dogs, often tinged with a gray-green pallor."[14] These are the horrid school lunches that have etched themselves into the public imagination. The situation got so dire in the wake of Reaganomics that McDonald's even offered to provide school lunches in exchange for advertising. In the 1990s, schools that contracted with the burger giant flew flags of the Golden Arches right underneath the stars and stripes on their flagpoles.

Meanwhile, feeding only low-income students, while ostensibly addressing individual food insecurity exactly as the NSLP intends, creates other problems, including an administrative burden on school administrators to chase down parents over unpaid school lunch debts. (Frankly, just reading the words "unpaid school lunch debts" should be a red flag that something is wrong in this country.) Requiring families to prove they are low-income before they can receive meals, called "means testing," can both harm students and undercut the effectiveness of the program. Being visibly branded by one's peers as a "free-lunch" recipient is stigmatizing and is often harmful to healthy development during childhood and adolescence, so much so that studies show that some students forgo free meals, even if they qualify, even if their parents pay, and even if they have to stay hungry. Former New York City Council member Ben Kallos, a school lunch recipient in his youth, has said the stigma meant that he had "to choose between friends and food."[15] This is a choice that no child should ever have to make.

Efforts to make school lunches more broadly available and more nutritious found a powerful champion in then First Lady Michelle Obama and her Let's Move! campaign. She was instrumental in pushing for the passage of the Healthy, Hunger-Free Kids Act of 2010, which provided additional federal funds for disbursement in schools whose lunches met improved nutritional targets. It has been invaluable in the fight against child food insecurity. Although school lunch is primarily a policy to address food insecurity, modern school lunches like those at PS56 in New York City adhere to the USDA's MyPlate nutrition guidelines, meaning that they're often healthier than what even wealthy students might otherwise eat.

Surprised? You may have read scare stories about school lunches of nothing but fried chicken and fries, but even when chicken tenders are in the mix at PS56, the plate is balanced and nutritious. Schools cannot get USDA reimbursement if they don't serve at least three key food groups (from fruits, veggies, grain, protein, and dairy) on each student's plate. In fact, a 2021 study published in the *Journal of the American Medical Association* of trends in US nutrition singled out schools as providing "the best mean diet quality of major US food sources."[16] During our visit, a gaggle of fifth graders tells us that chicken tenders are, indeed, by far their favorite lunch food. But if kids enjoy a moderate serving of less-than-healthy chicken fingers, they're more likely to eat the rest of the good stuff on the plate, resulting in fuller bellies and better nutrition. During our visit, the students devour grilled white-bread-and-cheese sandwiches, but they also polish off the beans, rice, and salad.

But while both supporters and critics of the Obamas latched on to the push for healthier school meals, it was a lesser-known part of the Healthy, Hunger-Free Kids Act that was the real "game changer," Crystal Fitzsimmons, director of Child Nutrition Programs and Policy at the Food Research and Action Center (FRAC), told us. The act included the Community Eligibility Provision (CEP) that allows school districts nationwide to provide free lunches to all students when a certain percentage of students qualify for free school lunches (40 percent when the policy was launched, updated to 25 percent in 2023). CEP "changes the whole culture of the school," Fitzsimmons said. Lunch goes from being a means-tested administrative burden and a potential source of social stigma to being just a taken-for-granted part of the school experience shared by all students.

CEP explicitly makes school lunches universal in poor districts, but it also reduces the cost to cities and states of shifting to universal school meals. Once again: Food access is a budget issue. If a large number of school districts already have their meal costs covered by the federal government, the potential cost of expanding to include all other districts—to make the meals truly universal—drops into the realm of fiscal and political possibility. In New York City, the largest school district in the United States, CEP covered schools attended by almost 75 percent of all students. That let the city make the leap to cover the cost of the other 25 percent under Mayor Bill de Blasio in 2017, offering universal breakfast and lunch to all students in all public schools. The program was launched, in a tasteful nod to Mabel Hyde Kitteridge, at a public school in Hell's Kitchen, realizing her dream of no child going hungry at school a century later.[17]

After securing funding and making programs universal, there remain two more pieces of the puzzle: working out the logistics and making sure that the food itself actually tastes good. In New York City, each school has a kitchen that prepares all meals for the day, but decisions about crafting menus and buying ingredients are made by the board of education. The menu options, which change every day and every quarter, are designed to meet the USDA's nutrition and budget targets. The results are pretty impressive, with high school lunch offerings ranging from that old standby pizza to chickpea shawarma with a side of curry potatoes or butternut-squash mac and cheese with crispy broccoli and fresh-baked cornbread. The city has also implemented Meatless Mondays and what it calls Plant-Powered Fridays to encourage more environmentally sustainable eating.

Crafting these menus requires democratic engagement. The city hosts tastings of new menu options with students and fields their feedback. Many new recipes don't make the cut. The ones that do are more likely to be eaten. "Rajma curry is a favorite" among students, Christopher Tricorico tells us of the Indian kidney bean dish. Tricorico, a former school principal, is the executive director of the New York City Department of Education's Office of Food and Nutrition Services. It's a title long and formal enough to sound like an army rank, which is fitting: Cooking 900,000 meals (breakfast and lunch) every day is a military-scale logistical operation. The city works with three large distributors that buy, store, and deliver all ingredients to every school in the city. Cooking the meals is the responsibility of 9,000 staff members, including cooks and assistant cooks and lunch workers at the schools and managerial staff at school, borough, and city levels. Getting all these moving pieces to work together to deliver lunches for $4 per student per day and breakfasts for $3 per day, day in and day out, is a small miracle.

Tricorico describes running the program as equal parts bureaucracy and catering. "We are in the business of customer service," he tells us. It's a sound-bite line designed to appeal to parents and reporters. But when you witness a working New York City lunchroom, it's pretty convincing.

At PS56 in the Bronx, Maureen O'Neill, the school's long-serving principal, provides countless smiles and high-fives to students during our visit. She has created a culture of not just joy but respect around the lunch program. Staff and students refer to all food workers as "chefs" rather than our own outdated "lunch ladies," and the chefs, in turn, work with impressive efficiency on tight schedules to deliver nutritious lunches.

It's not Le Bernardin or, for that matter, Zabar's. But for less than seven dollars per day, every child in New York City can be assured breakfast and lunch, taking strain off their families' budgets of money and time, and ensuring they get the most out of their education. And although the program has the predictable fiscally conservative critics, it also has critics who wish that it could do more. A chef who worked for years inside the program and has children in the school system—and who for those reasons asked to not be named—told us that at the end of the day, and despite all the good of universal school lunches, the iron rule governing school food is that "you can't betray your budget," meaning that ideals about nutrition and flavor sometimes get sidelined in favor of affordability. His solution? Increase the budget. "I wish American society put more value in nutritious food for everyone," he told us. Imagine what school chefs could do with eight, nine, or—dare we dream?—ten dollars per student per day.

Perhaps one of the few good things that came out of the COVID era was the temporary establishment of free school meals across the country. For a too-brief moment, we witnessed a historic victory for sound food-security policy. Follow-up studies confirm that it reduced stigma for students and reduced stress for parents, food secure and insecure alike. And a recent study of universal school lunch programs put some delicious frosting on this cake. It found that as kids eat meals at schools, families spend less at grocery stores. Grocery stores respond by reducing food prices . . . for everyone. In other words, universal school lunch programs are mildly counterinflationary.[18]

Alas, the nationwide free lunch program was discontinued at the end of the summer of 2022. But it did show several states that the benefits were worth the additional cost. Maine, Vermont, Massachusetts, Nevada, and California kept universal meals going starting that fall, and New Mexico, Colorado, and Minnesota have rolled out universal school meals since. Twenty other states have coalitions pursuing similar policies. Meanwhile, Senator Bernie Sanders, Democrat from Vermont, and Representative Ilhan Omar, Democrat from Minnesota, have introduced a bill to create a federally funded universal free school lunch program with a more generous reimbursement formula. Any politician who wants to credibly claim to care about the future of American families and children should cosponsor this bill.

IF NEW YORK CITY IS THE COUNTRY'S FOODIE STALWART, THE INKED-UP and cynical veteran with scarred and callused fingers that bangs out delicacies by force of habit in between smoke breaks, Durham, North Carolina, is the New South's foodie debutante, fresh-faced, eager to please, and too cutesy by half. The city's transformation into a cultural and economic hotspot has been rapid, buoyed by its culinary scene. *Bon Appétit* magazine has called it "America's foodiest small town."[19]

One sweltering summer day more than a decade ago, Gabriel, newly relocated to Durham from Indianapolis, hiked across the city and stumbled upon something that would shape his plate and palate for the next few years. On his way to meet friends to watch a game at the minor league baseball park, he cut across a swath of run-down blocks just north of the then-desolate downtown, a stretch filled with abandoned storefronts, weed-fractured parking lots, and a solitary craft brewery that had taken root in a former warehouse. But smack dab in the middle of that urban wasteland, he discovered a bustling farmers' market in the city's Central Park.

You can still find the Durham Farmers' Market, now in its twenty-sixth year of operation, in Central Park every Saturday, but the neighborhood around it is nearly unrecognizable. Some of the warehouses have been torn down, and others have been converted into apartments, breweries, hip coffee shops, farm-to-fork eateries, and even a sprawling food hall. A ritzy new building called the Vega, where a two-bedroom condo with views of the market will run you $1.5 million, has opened up just across from the park.

When Gabriel moved there, in 2011, the city's population totaled around 225,000 people; today it's swelled past 300,000. The hotter Durham (and its food scene) has gotten, the more challenging it has become for locals, including the people who work in its celebrated food and hospitality industry, to afford to live there. Housing prices have shot up, especially in the urban core where the farmers' market still sits. The average house lists for more than a half million. Foodie pleasures have proliferated, but so has food insecurity.[20]

The Durham Farmers' Market not only is a source of delectable produce and engrossing food chatter but also, to its credit, takes food insecurity seriously. Its website claims that it "is committed to increasing access to local and fresh food for ALL families and individuals in our community." Many of the vendors at the market accept SNAP. The market also gives customers additional coupons worth twice every SNAP dollar they use that can be spent on fresh produce and other programs for customers who have exhausted their SNAP benefits but still need food.[21]

But the sort of food—and foodie—culture of which urban farmers' markets are a part can, despite the best of intentions, create perverse and unintended consequences for food access. In an open real-estate market where, by definition, buyers bid against one another for scarce housing in desirable areas, amenities like a farmers' market, indie coffee shops, and farm-to-table eateries can increase housing prices and help displace low-income residents, hindering rather than helping with food access.

Given how closely linked malnutrition and obesity are with food insecurity, public discussions of food insecurity often tend to focus on the former rather than food insecurity itself. That is to say they ignore *access* (can people get the food that's available?) by focusing instead on the *availability* of food, and specifically of healthy options. Or they conflate the two. If low-income households had the same options available for purchase as affluent households, this thinking goes, they'd eat healthier diets—what the academic literature calls narrowing nutritional inequality. This idea has crystallized in the pop-scientific concept of food deserts: those neighborhoods that lack groceries, food stores, and other retailers of fresh food—a problem of availability. This theory pins the blame largely on *supply* factors, not *demand*, and it posits that increasing the supply of healthy options will improve food access and security for low-income households. In other words, put a farmers' market in an area that doesn't have fresh food, like downtown Durham circa 2011, and see nutrition improve and food insecurity decrease.

Now, we are strongly in favor of the availability of healthy options, but we're not sold on the argument that this inherently addresses the problem of food insecurity.

The idea of food deserts is a seductive and influential one that has wound up guiding—or misguiding—quite a bit of food writing and food policy in the past two decades.

In rural and remote communities, scarce options can contribute to food insecurity; in many small towns, Dollar Generals and gas stations may be the closest thing to a grocery. This is in large part caused by the lack of enforcement of an obscure piece of legislation called the Robinson-Patman Act, which for decades enforced fair competition in the grocery sector by prohibiting wholesalers from charging different prices to different grocery stores. For instance, it means that a wholesaler isn't allowed to charge less for broccoli or a can of beans to a massive grocery chain like Safeway than it does to a smaller chain or independent grocer. Nonenforcement means that large chains win on both sides of their business: They can afford to charge customers lower prices because of economies of scale, but they can also convince their providers to charge them less, further undercutting competition. Small towns have seen grocery stores shutter, only to be replaced by massive regional Walmarts or Safeways. For these rural shoppers, the choice is either to drive to the huge grocer or settle for meager local offerings.[22] In this sense, the rural food desert can be both a logistical and financial impediment to accessing healthful foods.

But the dynamic is quite different in cities like New York or Durham, where there's little evidence that food deserts drive food insecurity.

In urban and peri-urban areas, food deserts tend to have only a minor effect on household nutrition and diet. For the small number of consumers who lack reliable transportation, a new grocery or farmers' market could expand options—but only if it sold food at affordable prices. But most consumers, even most low-income consumers, don't shop at the closest grocery store. Rather, they travel to the grocery store that balances what they want to eat with what they can afford to eat. That is, most consumers are willing to spend the time driving, biking, or taking a bus an extra few miles to their preferred grocer even if there's a closer one just around the block. To be sure, some of this relates to the built environment of American cities, where wide roads and shoddy sidewalks make even close grocery options a forbidding journey for pedestrians, and housing and urban design policies that would result in more affordable and walkable built environments could alter this.[23] But a recent study in the prestigious *Quarterly Journal of Economics* put it this way: Supply factors explained only about 9 percent of the relative nutritional inequality between affluent and low-income households. The remaining 91 percent came from demand factors—that is, differences in what the households wanted to eat and not what they could access.

We praised Michelle Obama's support for school lunch programs, but now we must criticize the Let's Move! campaign's goal of eliminating food deserts. In support of that goal, rather than simply enforcing Robinson-Patman, the Obama administration created the Healthy Food Financing Initiative (HFFI), which provides grants to organizations and communities seeking to expand food options in underserved communities. Since its creation, the program, which like so many other food programs ended up housed in the USDA, has dispersed about $320 million in grants supporting "nearly 1,000 grocery and other healthy food retail projects in more than 48 states across the country." It has been renewed and expanded in subsequent farm bills and, in 2024, had authority to issue about $60 million in grants every year. State and municipal governments, as well as NGOs and philanthropic foundations, have also joined the fight against food deserts. To the extent that HFFI programs expand availability in rural and remote communities, they may have merit, but they frequently direct resources to programs focused on urban areas, where the justification is weaker and efforts to expand food options may actually be counterproductive.[24]

The logic of the HFFI and similar programs is that low-income consumers who reside in urban and suburban food deserts can't find nutritious fresh food—it is not available where those populations shop—and, as a result, they tend to have less healthy diets. If that's true, one way to improve food security is to subsidize the creation of fresh-food options in neighborhoods that lack them. Superstores and large corporate grocers such as Walmart, Target, Kroger, and Publix, of course, all have extensive fresh (albeit seldom local) produce offerings. But efforts to eradicate food deserts often subsidize retailers that have, at least rhetorically, a more community-oriented vibe: independent grocery stores with extensive produce sections, farmers' markets, community-supported agriculture, and farm-direct sales programs. This food feels good. But it is inherently more expensive than food that is sourced from conventional global supply chains.

Boosters claim that proximity to those foodie amenities does, however, lead to better diet and health outcomes. But does it? One influential study by economist Matthew Salois found that a well-developed "'local' food economy," including a high density of farmers' markets, "ha[d] a negative association with obesity and diabetes." If this were a causal relationship, then HFFI was right. Alas, the correlation fell short of causation: "Healthier communities [may] support alternative food systems which encourage a stronger 'local' food economy rather than the other direction."[25]

What Salois meant was that it didn't necessarily follow that farmers' markets made people skinnier; skinnier people might, instead, prefer to live around and shop at farmers' markets. In the United States, obesity is inversely correlated with income. Common sense and subsequent peer-reviewed literature suggest that boutique grocers and farmers' markets probably don't *produce* skinny people. Rather, they *attract* wealthier people who, on average, tend to be skinnier for a host of complex reasons that are not reducible to the store where they buy their kale and manuka honey.

Now consider the flip side of that observation: What if "food deserts" *attracted* lower-income residents, people who happened to have higher rates of obesity for other reasons, because those environments are often cheaper places to live with worse amenities? The consequence of watering food deserts with publicly subsidized food amenities might not improve food access in those neighborhoods—at least, not for very long. Instead, underserved neighborhoods would become more attractive to affluent consumers, demand for housing there would increase, and, ultimately, so would housing costs. This has been the case in downtown Durham. In time—and way less time than you might imagine—housing-affordability pressures would cause richer and skinnier food-secure residents to replace poorer, more obese residents who are more likely to be food insecure. A superficial gloss of those demographic changes would show that the neighborhood's residents had lost weight, but that would be a perverse statistical artifact of gentrification, not evidence of improved health outcomes *in the same population*.

A quick perusal of Zillow in your own community will show you that realtors believe that access to good groceries is a prized residential amenity worth advertising in their listings. This belief is backed by research. For example, a study of Washington, DC, found that a new grocery-store opening boosted average apartment rent in its vicinity by around 5 percent. This effect was strongest if the store was—prepare to be shocked—a Whole Foods, which garnered an 8.4 percent premium above area averages.[26]

Ironically, as we mentioned above, the retailers that do the best job of bolstering access to fresh, nutritious foods for low-income consumers are the big-box stores, none more so than Walmart. Walmart's ability to source food items from a global network of low-cost producers and to purchase in huge quantities gives it price advantages over many of its competitors. To be sure, this business model can be exploitative of producers, workers, consumers, communities, and sometimes competitors, and we are not advocating for Walmart as our grocery store ideal. But Walmart does deliver the most affordable fruits and vegetables to low-income households. A 2011 study found that uniformly lower grocery bills meant that Walmart's customers purchased and consumed more fresh and nutritious food simply because of lower prices, even as consumer-demand factors—not what was *available*, but what was *wanted*—determined the final quality of customers' diets.[27] Similarly, a 2019 study found that "closer proximity to a Walmart Supercenter improves household and child food security."[28] This, in turn, bears out the most basic fact about access: It's the price of food that is its biggest determinant.

When it comes to getting hungry people fed, it's like the Wu-Tang Clan put it many years ago: cash rules everything around me. Giving food-insecure people more money or free food or both does what watering food deserts cannot: It offers them a direct and expedient path to a better diet and a better quality of life.

IT'S A BRIGHT AND SUNNY SUMMER AFTERNOON IN DURHAM, AND we're eating churros with Nida Allam in front of Cocoa Cinnamon, a celebrated independent coffee shop. Allam is the chair of Durham County's Board of Commissioners, the highest elected position in county government, and this is where she likes to hold court—"the one with the churros," she texted when we set the location for the meetup.

Cocoa Cinnamon is an avatar of Durham's thriving food culture. It started as a bicycle cart at the Farmers' Market before opening its first brick-and-mortar location nearby. It's grown into a hugely successful business that delivers a quality product and has earned national acclaim and publicity. And Cocoa Cinnamon, along with its roastery Little Waves, is also mission driven. It aspires to be a paragon of ethical, community-engaged business practices. It sources fair-trade beans. It uses energy-efficient roasters and mostly compostable packaging. It pays a living wage, $17.60 plus tips. Its coffee is expensive by Durham standards, but its relatively well-off customers are willing to pay a premium.

Like many American cities, Durham's economy went through a very rough stretch in the second half of the twentieth century that saw its downtown abandoned. But the last two decades have witnessed a revival. Duke, the University of North Carolina, North Carolina State, and the nearby Research Triangle Park have all helped make the area a major biotech hub, and comparatively cheap space in downtown Durham made it an ideal home for both tech startups and the young, well-educated professionals they sought to hire. As a redevelopment strategy, the city's own marketing and PR material have helped to nurture its now nationally celebrated food scene, a jewel in Durham's crown that helps to draw and retain the city's affluent workforce. Those are, for the most part, the people getting caffeinated at Cocoa Cinnamon.

At only thirty, Allam is a rising star in Durham's political scene. She lives in the far east part of the city, an area that has seen rapid and arguably dubiously planned development, with tract housing and cheaply built apartment complexes outpacing infrastructure and amenities. Still, the abundant housing stock has kept down residential costs and kept the area racially and socioeconomically diverse. When we ask Allam what her favorite restaurant in Durham is, she points to a modest halal pizzeria in a strip mall near her house whose proprietors lovingly fed her during a recent pregnancy.

That part of the city is also home to many of the Durhamites left behind by the city's success story. Eighteen percent of Durham residents and 20 percent of its children are food insecure, with the city's Black and brown residents most afflicted—one in six Black workers and one in eight Latino workers in Durham have reported skipping meals to save money. It's part of Allam's job to address this problem. Several programs, including community gardens, food hubs, and meal-assistance providers coordinate through the county's USDA extension office, which often connects community partners, for-profit and not-for-profit alike, to federal grants and useful information.[29]

But as with New York City, the most potential for a big food-security win is in the schools. Even as Durham has grown noticeably richer and whiter in the past two decades, its public schools have grown poorer, browner, and blacker. In response, Durham Public Schools (DPS) launched a universal breakfast and lunch program for its thirty thousand students in 2024.[30] It's a big victory, but with asterisks. Purchasing guidelines obligate DPS to source from approved vendors over local vendors they'd prefer, Allam notes, so students sometimes wind up with "syrupy fruit cups instead of freshly cut fruit."

But the bigger problem is with the professional cooks who make the meals. School Nutrition Services staff takes pride in feeding DPS's students. But they're also workers with their own household budgets and housing costs, and DPS has struggled to pay its employees wages that can match the rapidly rising costs of living in Durham. DPS's struggle to both feed its students and pay its workers fairly crystallizes a bigger issue. There are many places to eat in Durham, from joints that get glowing name-drops in *The New York Times* to the less vaunted Waffle House on Apex Highway. They all depend upon the labor of workers who, increasingly, cannot afford to live in Durham.

Patrick, a native of Tennessee, has worked in food, drink, and hospitality since he arrived in Durham eleven years ago, and his views echo the sentiment of many workers with whom we spoke. He's tended bar, waited tables, catered, and been a barista and (eventually) manager at a coffee shop. He currently has two jobs: one managing a nightclub and another bartending, serving, and managing at a farm-to-table-style restaurant a few blocks from the farmers' market. The restaurant captures the local culinary spirit. "Local ingredients, focused on sustainability. . . . Dogs are welcome. Very Durham," Patrick laughs.

The restaurant was also one of the earliest businesses heralding the Central Park District's renaissance. "The area used to be known as the DIY district," explains Patrick, "because there were food trucks that would pop up there . . . and they would turn into brick-and-mortar shops." Low rent and ample opportunities to start small businesses drew incredible cooking and hospitality talent—people could afford to live, work, and play in the neighborhood.

That spirit has become frayed by the area's development, which has increasingly driven a wedge between affluent customers and the service workers scraping to get by.

To stay relatively close to his two jobs, Patrick has jumped from lease to lease, and he has still seen his rent nearly double in the past five years, a down payment on a condo always out of reach. Rent in the Durham–Chapel Hill metropolitan area climbed by about 35 percent between 2018 and 2022, and median house prices leaped 75 percent. Housing affordability has a major impact on day-to-day food security for restaurant workers. More money for rent is less for food. Patrick has plenty of friends who take SNAP benefits. Shift meals provided by their workplaces are, for some, literally the difference between being sated or going hungry.[31]

But the real tragedy is that like most of the Durham restaurant workers we spoke with, Patrick derives meaning and joy from the job. He loves the community. He loves the food. He loves Durham. And he wishes local government and real-estate developers would keep all that in mind as they sketch out the city's future. He'd like to see programs in the mammoth new complexes that offer rental breaks to people who work in businesses in the area. That may not square with short-term profit, but if housing becomes unaffordable for restaurant workers, it will kill the goose that laid Durham's golden foodie egg.

City and county officials are aware that food security and housing affordability are interwoven issues. Best practices converge on planning that keeps that in mind, leveraging nonfood policy to ease food insecurity and using food policy to assist housing affordability. Durham's Housing Authority, for example, has several ambitious public-housing projects underway thanks to a $95 million bond issue approved in 2019.[32] Several of these projects are within walking distance of downtown and may improve affordability for workers like Patrick. The largest of these is being designed as a mixed-income affordable-housing community with retail space for a grocer at ground level. The city can sometimes use the zoning process to force developers to include affordable housing in their complexes, but only in some cases. And, even then, some city officials caution that driving too hard of a bargain with developers can stall projects and depress the city's overall housing supply in ways that drive rents up further still. But policies that expand affordable housing options are likely to be Durham's best bets for food workers' food security . . . and for keeping its vibrant food culture thriving.

IT LACKS THE ROMANCE OF COMMUNITY GARDENS AND FARMERS' markets and the verve of pathbreaking chefs or entrepreneurs, but food access is primarily a budget problem. As such, transformative programs for low-income people look more like a debit card than a slow-food revolution. They involve food served in a school cafeteria or provided by a food bank, not a hot new menu item at a chic café. School lunches and SNAP benefits can ease food budgets without directly increasing housing costs. And, at scale, they rely on working with retailers in the conventional food system, be it through SNAP funds or the whole-sale buying done by school lunch programs and food banks. Increasing these programs' budgets will ensure more access to better, more nutritious meals for more people. The corollary of this is that conventional supermarkets tend to expand purchasing options for healthy and nutritious food (even if the option is not always taken). Likewise, policies that ease housing costs and boost incomes for working people can provide budgetary slack that helps them eat better.

Housing and eating are interwoven, as are wages and grocery bills, different sides of the same coin. So too are *eaters* and *workers*. If he waited on you at his restaurant, you might only consider Patrick as the latter, but everyone who works in a restaurant eats. As the next chapter shows, for many workers, better food, better wages, and better working conditions are related struggles. That means the greatest potential allies for many eaters striving for a better food system may just be the people serving them their meal.

Chapter 5

WORKERS AND EATERS

THE MERCURY IS TICKING NORTH OF 95 DEGREES, AND THAT'S before you adjust for humidity. We're standing across the street from a Waffle House in downtown Atlanta next to Centennial Park, hiding in the shade of a tree. Current and former Waffle House employees, as well as organizers from the Union of Southern Service Workers (USSW) and several allied unions, are converging here to make some noise about what they say are the abysmal working conditions in Atlanta's many Waffle Houses.

Waffle House, a place that's an accessible, delicious, and affordable part of the food landscape—even if a not entirely healthy one—is something else entirely for many of its workers. And they've decided to do something about it.

We chat. Cigarettes are smoked. Coffees are drunk. The energy is building.

Then we're out of the shade and marching on asphalt sticky from the Southern sun and surrounded by nearly fifty workers and organizers, nearly all Black and Latino, mostly women. Two young men brandish a sign that reads "Stronger Together," and that's the vibe: joyous and raucous solidarity. There are labor songs and familiar chants, including the fearless call-and-response of the USSW.

"Who are we?"

"THE USSW!"

"What do we do?"

"ORGANIZE THE SOUTH!"

Soon, Waffle House employees are speaking over a bullhorn about what they say is going on in their restaurants—the poverty pay, the dangerous working conditions, and the tin-eared executives who, for all their talk of the restaurant chain being a family, treat their employees with contempt. When workers delivered a stack of 450 worker-signed petitions outlining their complaints to the company's Atlanta headquarters, they say corporate literally tossed it in the garbage in front of them.

Twenty-two million Americans work in agriculture and food either full- or part-time, representing 10 percent of all American jobs. They feed us, yet they are the most overworked, underpaid, and undervalued members of the workforce. There are reams of statistics to prove this, but you can just listen to the workers themselves.[1]

Cindy, a veteran Waffle House employee, whom we talked with immediately before the protest, starts things off. Her voice is gravelly with a mild Georgia twang layered over the remains of a flat Midwestern accent. Cindy uses the cigarette she's smoking like a conductor's wand to add emphasis to her expressive hand motions. After working at Waffle Houses in Georgia for thirty years, she still earns less than $3 per hour before tips, a rate of pay so dreadful that she frequently struggles to afford groceries.

"How we feeling today?" Cindy asks the crowd over the bullhorn. "I feel good because I just delivered a demand letter to our store."

Poverty pay is common for Waffle House servers. The company maintains that with tipping included, servers wind up earning closer to $14 per hour. Many Waffle House servers roll their eyes at the tortured accounting that produces that number. Reporters who dug into this issue noted that "almost a quarter of the company's 42,473 employees take home between $2.13 and $10 an hour, in both wages and tips," and, regardless, even if it were $14 per hour, that's still short of a living wage: the minimum hourly wage necessary to afford all basic necessities like food, housing, health care, and transportation. According to MIT's Living Wage Calculator, a tool that estimates living wage based on national consumer spending data, the living wage in Atlanta is $24 per hour.[2]

After Cindy delivered the demand letter, a formal notification to an employer that their employees are attempting to form a collective bargaining unit, her restaurant's staff eventually went on strike. It was one of several actions organized by the USSW that has begun to wring some concessions from corporate. They've agreed to a $5.25-per-hour base wage to be implemented in June of 2026 followed by a boost up to $7.25 per hour in 2027. The workers say that's neither good enough nor soon enough. They're demanding $25.

Low pay is the biggest complaint, but it's far from the only problem. Workers are also angry about what they describe as the deteriorating and unsafe working conditions that have been allowed to fester at many of Waffle House's 1,900 locations. Equipment and furnishings go unrepaired for weeks and months. On sweltering days like today, they hope and pray that the air conditioning holds on. When a location in South Carolina lost its AC in the summer of 2023 in 100-plus degree heat, outraged workers went on strike in protest.[3]

And then there's the fighting, especially late at night. Waffle Houses have always attracted drunk customers when the bars let out, an unavoidable aspect of being open 24/7. But lately that crowd has been getting rowdier. Customers brawling, threatening, and hurling abuse (and furniture) at staff is a phenomenon that has become so common that it is a recognizable genre of viral content on social media. The behavior—and the virality—has created a negative feedback loop: Troublemakers scare off the folks who just want a late-night waffle without a side of trouble. Some Waffle Houses have responded by closing the indoor seating area at night and routing everything through a to-go walk-up window instead, but that just turns the parking lots into the Wild West. "You can still stick a gun through a window," Cindy, who's seen it all, reminds us. Meanwhile, servers and fry cooks do unwaged work as a self-defense force—viral videos of Waffle House employees whupping ass suggest they are formidable fighters—but management and corporate doesn't pick up the bill if they're injured defending themselves.

As if all that wasn't enough, there's also the "meal deduction" program. Waffle House deducts the cost of a meal—they pin it at $3.15, or more than Cindy makes in an hour—from the paychecks of all workers every shift. That winds up being about $30 million in cash that Waffle House harvests from its own workers every year. Although workers are indeed entitled to claim a meal, including theoretically making one worth far more than $3.15, the deduction is compulsory and automatic, meaning that workers pay for it even when they don't eat it. And workers may opt not to claim the meal for a variety of reasons. They may be too busy to eat, have dietary restrictions, be worried about the long-term health consequences of eating too much of the food that's served at Waffle House, or just want to eat something else sometimes.[4]

Now, you might come to some hasty conclusions listening to these complaints, which make the Waffle House sound less like a great place to eat and more like a depraved Roman coliseum of dining where the gladiators serve you hash browns before they fight. You might assume that Waffle House employees don't like Waffle House, don't like the food they serve, and don't like the work they do. You might think they'd like to raze all Waffle Houses and opt out of the food system they represent.

And you'd be wrong.

For starters, the folks at the protest we talk to love the food. Everyone has a favorite order, reflected in the intricate, back-catalog, and off-the-official-menu delights they mention: things like the "Texas Grilled Chicken Melt" and the "Cheesesteak Melt Hash Brown Bowl." One woman lists out a personalized sandwich order so lengthy and complicated we can't jot it down in time before adding "and a waffle. You've gotta have a waffle."

And as justifiably furious as Waffle House's workers are, they are organizing because they want their workplaces *to improve, not to disappear.* This is why they loft a banner that reads "Waffle House: Our Dedication Deserves Dignity."

Just as Waffle House workers' complaints give us a candid look at the restaurant's failures, their demands also represent a goal *and* the means to achieve it. Empowering them to define the conditions of their own labor can mean better pay *and* a solid constituency that cares powerfully about improving the food system. In mid-century America, unions successfully made the case that wages should be good enough that workers could also afford to be customers. We'd now suggest a similar goal for today's food system: The millions of people working to feed everyone should have dignified work that would pay them enough that they could afford to eat well.

FLASHBACK: IT'S THE SPRING OF 2012. GOTYE'S "SOMEBODY THAT I Used to Know" and Carly Rae Jepson's "Call Me Maybe" are on top of the charts. President Barack Obama is locked in a tight reelection race against Massachusetts Governor Mitt Romney. Amid a general election that is just beginning to heat up, the Department of Labor takes quiet action on a seldom-discussed problem: It makes some modest revisions to the rules that regulate how children can be employed on American farms.

Then, as now, hundreds of thousands of children labored in the United States to put meat and fresh dairy, fruit, and vegetables on dinner plates. And we don't mean kids setting the table or chopping carrots for a family meal. The federal law regulating the employment of minors in the United States is called the Fair Labor Standards Act (FLSA). But the FLSA rules that constrain most other employers, from car washes and cinemas to warehouses and factories, simply don't apply to farms. In those other workplaces, children under eighteen are prohibited from operating heavy machinery and working at dangerous heights, and fourteen is the minimum age for most jobs. Not so on farms. Farmers enjoy a unique set of exemptions to the FLSA (and many state-level laws) that functionally permit children to work at much younger ages doing far more dangerous jobs. Federal minimum labor age in farming drops to twelve with parental permission. In Oregon, children as young as nine are legally allowed to work picking berries and beans. In Hawaii, the minimum age for coffee harvesting is ten. In Illinois, it's ten regardless of the type of farm work done. In Utah, there is no minimum labor age if parental consent is given. Reliable estimates of child labor in US agriculture put the number as high as 800,000.[5]

Every few years over the past half century, the media have run hard-to-stomach exposés and op-eds about these kids. Back in 1981, *The New York Times* published a particularly jarring account by environmental-writing stalwart Paula DiPerna that began with an eight-year-old harvesting asparagus in eastern Washington state with the keen eye and practiced routine of an old farmhand.[6] There was the widely publicized case of María Isavel Vásquez Jiménez, a pregnant seventeen-year-old who died from heat stroke after pruning grapevines in the California Central Valley's triple-degree heat for nine hours in 2008. Media coverage and subsequent public outrage about Jiménez's case galvanized some changes to California's child labor laws.[7] But the case of Glen Nolt and his two sons, age eighteen and fourteen, all of whom drowned in a cow-manure lagoon on a large Maryland dairy farm in 2012, is more typical: The media lamented the tragedy but didn't ask why children were working around a manure lagoon in the first place.[8] And then there are cases like the fifteen-year-old boy in Guthrie, Texas, who was crushed under heavy machinery while installing fencing in 2022. His death led to a $20,000 OSHA fine for his employer, but his status as a minor in a remote rural community kept his name out of the news.[9] And these sorts of preventable deaths sit next to far more common nonfatal accidents, most of which don't make the news at all but can result in lifelong disabilities.

The National Children's Center for Rural and Agricultural Health and Safety monitors statistics about child labor, injury, and mortality, and its findings are sobering. A child dies in American agriculture, on average, once every three days, and thirty-three other young workers are seriously injured every day. Between 2001 to 2015, 48 percent of *all* fatal injuries to under-eighteen workers occurred in agriculture.[10]

Children have been working in American agriculture as long as there have been farms. A century ago, most did so on their family's farm. The situation has grown more complex since then as farms have grown larger and become increasingly reliant on migrant contract and wage laborers. Children in agriculture often still work alongside their families, but that's because many are the children of wage laborers, many of them migrants. In many of these families, children either need to work so the family can survive, or the families themselves face retaliation from their bosses if they don't pressure their children into the fields.

Worse yet, *legal* child labor also camouflages *illegal* child labor. Regulations with so many holes already look like Swiss cheese, but one more provision melts them into an unenforceable goo: Child workers in agriculture are not required by the FLSA to provide proof of their ages, which means that farmers could comfortably look the other way and plead ignorance if anyone noticed, say, eight-year-olds harvesting asparagus.

The idea is repulsive to most normal people, which is perhaps why, in 2012, Secretary of Labor Hilda Solis confidently invoked the discretion granted by the FLSA and announced revisions to the rules so that farmers had to operate by the same child labor standards as everyone else. Anticipating that some family farmers might (not entirely unreasonably) complain that the rules encroached on their freedom to parent how they liked, the revised rules maintained exceptions for small farms where the child was related to the employer. In other words, the revised rules were narrowly tailored to target hotspots for the most serious exploitation: large produce, tobacco, and dairy farms. It was one of the most commonsense regulations one could ask for. A moral no-brainer supported by the overwhelming majority of Americans.[11]

The pushback from farmers and their political allies was swift, brutal, and effective. The American Farm Bureau Federation, the major lobbying force for American agribusinesses, along with its state-level affiliates, launched a broadside against what they claimed was the Obama administration's plan to ban "farm chores" and 4-H clubs. Farm chores and 4-H clubs? The target of the regulations could hardly be defended, so the farm lobby distorted the intended changes, making small family farmers, *those explicitly exempted from the new regulations*, out to be the victims of oppressive government overreach. This critique was echoed by Republicans like Sarah Palin and media outlets like *Fox News*. But it was also the line taken by Democratic politicians from farm states, such as senators Al Franken of Minnesota and Jon Tester of Montana. Even more shockingly, there was no organized pressure from progressive political interests—or, for that matter, the left-leaning writers and activists who usually expound about food justice—in support of these regulations. The myth of honorable family farmers and ennobling farm work carried the day. With pressure coming from Democrats in crucial swing states in an election year, Obama's Department of Labor scrapped the planned revisions.[12]

The loopholes identified back in 2012 remain in place. In the summer of 2023, a collection of progressive legislators introduced the Children's Act for Responsible Employment and Farm Safety to address the situation. The bill's author, California Democrat Lucille Roybal-Allard, has been introducing similar legislation since 2005. It went nowhere then, and it went nowhere in 2023. The statistics on child labor have barely shifted from 1981 to 2012 to today. The stories behind those statistics are as heartrending as ever.[13]

The grinding inaction on child labor in American farming powerfully illustrates who wields power in contemporary farming and who does not. Eliminating child labor—child labor!—should be low-hanging fruit for anyone interested in practically and justly changing the food system. Of course, talk to farmers or the lobby groups that represent them or the legislators who want their votes, and child labor is either denied outright, explained away as a practice engaged in by bad apples, justified based on the seasonality of harvests that often fall over the summer school break, or shrugged away as a minor problem. The statistics and the furious lobbying of farmer and agribusiness groups tell a different story: Child laborers are cheap, easy to coerce, and expendable. And though it's difficult to swallow, those qualities make the child labor problem a microcosm of how labor more generally is treated in the American food system.

THE SIMPLEST WAY TO BREAK DOWN LABOR IN THE FOOD VALUE CHAIN is to divide it into five categories: production, processing, service, retail, and distribution and delivery. Respectively, these employ 2.6 million, 2 million, 13 million, 3 million, and 1.4 million people, a total of 22 million workers.[14] Take that Waffle House waffle, our ongoing exemplar of delicious, affordable, mass-produced food. For it to get to your plate, someone had to plow the soil and plant the seeds on one of the 100,000 or so farms in the country that grow wheat, likely with a large, motorized sower. Someone had to harvest it. Someone had to transfer the grain onto trucks or trains to get to grain elevators. There someone had to load it onto more trucks or trains to get it to a flour mill. There someone, using heavy machinery, had to clean it, blend it, grind it, and package it. Then someone had to yet again get it all onto trucks to deliver to Waffle Houses. There, back-of-house staff had to unload it, cooks cook it, and front-of-house staff serve it. If you're wondering about the premade waffle mix you can buy off the Waffle House website or at a store, someone has to blend that too, and someone else at a grocery store has to put it on a shelf and check you out. Every one of the hundreds of people throughout this value chain—with potentially the exception of a farmer-operator driving a harvester—will be among the lowest-paid workers in the American economy. And the people making your waffle and serving it to you happen to be the worst paid. Fast-food cooks and servers make up one of the largest groups of employees in this country, but they average just around $30,000 annual income. In fact, eight out of the ten lowest-paying jobs in the country are food system jobs. The pay is so poor that the people who feed us are 50 percent more likely than the average American worker to use SNAP benefits and almost 100 percent more likely to be food insecure.[15]

How did we get here? The answer, to some extent, lies in the brute market forces of supply and demand. Most food-sector jobs have few educational prerequisites, and many do not require previous labor experience. These jobs tend to attract a young and immigrant workforce. This massively increases the number of people for whom working in food is an option, meaning that farms, warehouses, and restaurants can get away with setting low wages because they'll be able to fill the jobs.

But wages aren't determined only by the invisible hand of the market. They have a fixed floor. And business interests throughout the food system have for a long time been adamant about keeping that floor as low as possible, lobbying for decades against minimum-wage increases at both the federal and state levels, and in some cases seeking special exemptions, such as those for tipped workers, that drop the minimum wage far below the poverty line. The United States is many things, but it is not a country whose government has historically treated workers' welfare as a priority. Until 1966, many jobs in agriculture and tipped jobs in food service were excluded from the federal minimum wage law altogether. Many critics contend that this is because these jobs were overwhelmingly done by women and non-white men. When the administration of President Lyndon B. Johnson, facing pressure from labor organizers and the civil rights movement, expanded the minimum wage to many previously excluded jobs, it kept a proviso for tipped jobs, which set their minimum wage much lower on the assumption that tips would cover or beat the difference. If tips didn't meet a basic level, employers would have to cover the difference. The tipped minimum wage would periodically go up as the minimum wage went up, but it would never close the gap. If you've ever worked in food service, you know where this is going. This system opens workers up to tip theft and tip redistribution, and puts their take-home pay at the mercy of customers or bosses.

Tipping is a good metaphor for the food industry's labor strategy. It is, in effect, a private tax levied by restaurants on customers to subsidize their employees' wages. It's an extra charge on your meal, upheld not by the law but by the force of social pressure and human decency. The idea that a worker must both work their actual job and then work extra to earn a tip, including by bending over backward for unpleasant customers, is demeaning. On the other side, feeling compelled to spend an increasingly high percentage on tips compared to the past is an unpleasant and unfair imposition on diners. The debate about tipping is often framed in terms of either manners or respect for labor, but it's an artificial one that has been caused by the restaurant industry fighting against actually having to pay living wages. What's worse, the restaurant and fast-food industry often claims that raising wages or providing employees with benefits makes their businesses unfeasible, meaning that if they pay workers more or treat them better, there will be fewer jobs to go around.

Perhaps the most infamous defender of this status quo was Herman Cain, a contender for the Republican presidential nomination in 2012. Cain's claim to fame was that he was the former CEO of the National Restaurant Association, a restaurant and fast-food lobbying group. He had also worked for Burger King and Godfather's Pizza before taking over the NRA (same acronym, similar strategies, different product). As the head of the NRA, he spearheaded opposition to increases in the tipped minimum wage precisely on the idea that raising it from $2.13—a level it has not risen from since—would disincentivize restaurants from hiring staff. As if the food was going to serve itself. Throughout his career, the cartoonishly villainous Cain consistently pitted himself against the interests of working people based on the claim that the food industry simply couldn't afford to treat them better. He was completely wrong. We know this because we have a counterfactual example of what happens when food companies must adhere to much higher wages and benefits.[16]

Back in 1981, McDonald's wanted to bring its burgers to Denmark, in part to ensure that its rival Burger King, which had set up shop in the Nordic country in 1977, didn't corner the market for American fast food. Denmark has no minimum wage, but it does have an economic system predicated on expected (but not mandatory) collective bargaining between unions and employers at the company or sector level. Sectoral bargaining is a process whereby unions represent all workers in each sector, like food service, across the entire country, bargaining for wages and working conditions on behalf of a large number of workers doing similar jobs and therefore making similar demands of employers. Denmark already had—and continues to have—a robust welfare state providing universal health care and other benefits, so bargaining mostly concerns wages, holidays, and on-the-job safety and benefits.

Unlike Burger King, which had opted to play ball on Danish terms, McDonald's decided to set its own wages and holidays. In true American style, it ignored criticism and polite overtures from union representatives and politicians. After a few years, the Danes lost patience. The hotel and restaurant workers union staged protests of McDonald's and called for a boycott by consumers while more than a dozen other unions engaged in a solidarity strike: a strike in support of another union's grievance and not their own. Delivery drivers stopped delivering; bakers stopped baking buns; construction workers refused to build or repair. And just like that, McDonald's capitulated and agreed to the terms of the union contract. Today, McDonald's workers in Denmark make the equivalent of $20 per hour in base pay, get six weeks of paid vacation time off, and get a pension. Yes, a pension. And McDonald's continues to operate profitably in Denmark.[17]

The principal reason why McDonald's doesn't offer the same compensation package in the United States is because fast-food workers here lack bargaining power. Although some industries in the USA such as Hollywood screenwriting, airlines, and hotels are subject to sectoral collective bargaining—as in Denmark—the food industry isn't. Some food workers fall under hotel workers' unions. Others fit under umbrella groups like UNITE HERE! (which covers some workers at airports and in transportation) or the UCFW (which covers some workers in grocery stores and food processing). But most food service workers—upward of 98 percent—are not union members and have little political or legal support in their dealings with their employers. The result is low pay, few benefits, and little to no control for workers over basic working conditions like weekly schedules. Bosses are unlikely to give workers anything they don't fight for, so McDonald's and groups like the NRA have lobbied for decades to depress wages and benefits. They also fight tooth and nail against unionization and worker organization.

But the last few years have seen a surge in food labor organizing, ranging from the sorts of protests we saw in Atlanta to organized fights for increased wages to the establishment of something very close to the Danish sectoral-bargaining model for fast-food workers in California.

IT WAS NEW YORK CITY FAST-FOOD WORKERS WHO SET OFF THE CURrent wave of labor mobilization in the American food system when, in 2012, they walked off the job and demanded a $15-per-hour wage. The campaign was born from community meetings over skyrocketing rent prices in the city, and many of the people being priced out were food service workers. Minimum wage in the Big Apple was set at the federal minimum wage of $7.25. The workers who walked off the job asked for more than just a raise; they inspired a rallying cry for a living wage and better working conditions around the country.

At first, given their demands, they were ridiculed or, worse, ignored. But by organizing across cities and with the support of the Service Employees' International Union (SEIU), which took up their cause, they escalated. Few people remember it now, but in the fall of 2014, protest marches led by fast-food workers for better pay and more dignified working conditions raged through major cities. Hundreds of protesters were arrested. At the time, the NRA (the restaurant one) called the protests "orchestrated union PR events," which technically is correct, but not in the disparaging way they meant it.[18] With constant pressure and a growing number of supporters, especially progressives in city-, county-, and state-level legislatures, what was initially dismissed as utopian actually started happening. California and New York state, as well as New

York City itself, were quick to establish minimum wages of first $12, and then $15. Naysayers were soon either convinced or pressured. Former New York Governor Andrew Cuomo, who had previously called a $15-per-hour minimum wage a "nonstarter," was by 2015 leading a task force on state-level wage increase phase ins.[19] Today, just over a decade since the Fight for $15 was launched, the starting salary at a McDonald's in New York City is $16. Of course, the cost of living in the city has also risen precipitously, but the fact remains that workers willing to make big asks and fight for them can get the goods.

Food service is seemingly a far cry from the model of industrial manufacturing that spawned the modern labor movement and from the heavily unionized factories and slaughterhouses of the postwar era. But while these workers are wielding spatulas, aprons, and dishwashing gloves rather than wrenches, their workplaces are rippling with the potential of organizing every bit as much as the Detroit car-assembly lines. Like any workplace, a lot of that has to do with shared experience on the job, of shared concerns about safety or complaints about bosses, of worries about making it to pick kids up from school after a shift, of the uphill battle of saving up for college. These are the conversations had over hot grills or sitting on a curb on a smoke break. Notably, these are also workplaces—groceries, gas stations, and restaurants—that are relatively easy for union organizers to enter (and potentially picket) because they rely on members of the general public entering and interacting with their employees on a regular basis.

In the age of social media, stories of the common struggles of food service workers increasingly pop up online, like that of the Ohio Waffle House worker who took to TikTok to show that his reward for having been the grill man for $2 million worth of sales was a black company polo shirt emblazoned with gold lettering reading "2 Million Dollar Club."[20] Not a bonus like a top salesperson would get at a white-collar job, but a shirt to wear on the job to drive home just how much money the company had made off his labor. Minimum wage in Ohio at the time was $9.30 per hour. Sure, sure, you say, but it's the man's job to work the grill and drive sales. Absolutely, that is the purpose and nature of waged labor, but a shirt celebrating his revenue generation is just a little too on-Karl-Marx's-nose.

These stories speak to shared hardships that can build solidarity, and it was and continues to be the strength of the organizing in these spaces, often led by women and workers without higher education, that got the goods by putting their voices and interests on the agenda. But—and this is crucial—there is a big difference between winning a one-time wage increase and winning the wide range of protections and bargaining power that comes with a union.

The 2020s have seen a rise in union activity in food, with members of the Bakery, Confectionery, Tobacco Workers and Grain Millers' Union at Kellogg's, Nabisco, and Frito-Lay plants engaging in strikes and other labor actions to secure better contracts, including winning control over scheduling to prevent mandatory overtime or back-to-back shifts that not only make seeing family difficult but also increase the risk of fatigue and therefore on-the-job accidents. But while food service workers may have won some wage increases through the Fight for $15, creating unions has proven far more difficult, in large part because the large food service chains like Starbucks that employ hundreds of thousands of people and where unionization would do the most good have done everything in their power to prevent it from happening.

Starbucks is less a coffee shop than it is an empire that has grown wealthy off the hopeless addiction of the American people to caffeine and convenience. Nineteenth-century England had opium. The Starbucks Corporation has little roasted beans. In 2023 the company had a total global revenue of just under $36 billion. In the United States alone it runs 10,000 stores and employs more than 200,000 people who dedicate their days to serving up the extravagant, criminally overpriced, coffee-flavored, whipped-cream-topped milkshakes that the company has convinced some people is an acceptable desecration of a once-sacred beverage.[21]

Much of the incredible wealth earned by Starbucks does not go to its workers. While Starbucks workers had dabbled in unionization since the 1980s and had unsuccessfully attempted a unionization drive in the 2000s, the Fight for $15 and the rising tide of American labor organizing in the late twenty-teens galvanized a new generation of organizers set on unionizing Starbucks baristas.

Now let's pause for a second. Being a barista at Starbucks isn't the best job in the world, but as far as service-industry jobs go, it's not too shabby. It tends to pay slightly above minimum wage and offers a range of benefits, including health care and relatively generous parental leave, that far exceed those of McDonald's or Waffle House. But this wage is still often low, especially in major cities, and the point of unionization is as much about political power within the workplace as it is about dollars and cents. The United States is a democracy, but most people, and all waged food system workers, spend their working days in environments that are mostly autocratic. In the private sector, governed by contract, employers have almost complete control over workers' wages and work environments as long as what they do isn't literally illegal. If negotiations over everything from wages to workplace harassment to unfair dismissal are left to be settled between individual employees and their employers, the power differential means that individuals will almost never get their way. The collective power offered by union membership, including the power to negotiate as a bloc and to strike and withhold labor, is one way of balancing out power in the workplace and ensuring workers have a voice.

The Starbucks Union efforts started in one store in Buffalo in 2021, which voted to form a union that December.[22] Since then, the movement has grown steadily, and through steely determination, Starbucks

Union organizers have unionized over 11,000 workers across 481 stores in three years. And they have done this against a brutal counterattack by Starbucks, ranging from anti-union messaging to firing organizers to hiring union-busting law firms.[23] The company has not revealed how much money it has spent on its anti-union efforts, but estimates in a letter shareholders sent to the Federal Trade and Exchange Commission place the number at over $240 million.[24] It has also done things like increase pay and perks for nonunionized stores, which is not only incredibly petty but also technically illegal.[25]

Considering this, the organizing effort has been nothing short of heroic. But organizing in this way, store by store, is slow and cumbersome. In part, this is because of the nature of the national legal bureaucracy. The National Labor Relations Board (NLRB) must oversee and certify each union election and address complaints about it. Labor lawyer Matt Bruenig has argued that passing a piece of legislation called the Employee Free Choice Act could facilitate the process and make unionizing efforts like those at Starbucks easier. Proposed in different versions since 2003 but never passed, this legislation would allow workers to form unions if a majority in a workplace expressed their desire to join a union in writing (with no NLRB-supervised vote) and would then give employers ninety days to engage in bargaining or be forced into government-supervised mediation. Passing it would cut red tape and facilitate unionization.

Bosses hate the idea, which may explain why even when Democrats have held majorities in the House and Senate, they haven't passed it. But in the long term, legislation that makes organizing easier and gives organizers greater legal protections will be the best way to ensure that workers can participate in the politics of work. And as the case of Denmark shows, having this sort of participation does nothing to clog the gears of commerce; in fact, it lubricates them. Denmark regularly beats the United States on business competitiveness rankings despite having the very sort of worker protections and universal benefits that American businesses claim would render them uncompetitive.

Sectoral bargaining like that in Denmark has long been considered a pipe dream in American food service. But as of 2022, it is a reality for fast-food workers in California. Also growing out of the Fight for $15, California's food-service workers fought both for higher wages and for something akin to the Danish model. By not only staging strikes but also proactively getting legislators on their side, they got both. Minimum wage for food service in California is now $20, and the state passed Assembly Bill 257, which creates a Fast Food Council that can engage in bargaining on behalf of the state's 550,000 fast-food workers. Although the bill was slightly softened and renamed in the wake of challenges from the NRA and other corporate entities, the legislation is now a model for how workers can gain more political power to lobby for their interests.

CHRISTIAN SOSA (NOT HIS REAL NAME) REMEMBERS HATING THE TERM "essential worker" as soon as he started hearing it during the pandemic. It had a bitter taste to it that he choked on every time a restaurant wouldn't let him use the restroom or charge his phone or electric bike battery, or even eat inside, even when it was raining.

"'Essential' means *indispensable*, right? Well, it didn't feel that way," he tells us in Spanish. If anything, he felt highly dispensable.

At the height of the COVID outbreak in New York City, there were more delivery cyclists on the streets than cars, ferrying groceries, medicine, and takeout food to sick, locked-in, and scared New Yorkers. Sosa says that keeping the city fed didn't feel uplifting when he was pulling up his helmet and pulling down his mask to scarf a slice of pizza on the sidewalk between deliveries. In those days, his phone would beep nonstop with orders piling up, giving him little time to even make deliveries on time, much less eat healthy meals. Arriving late or missing orders would lower his rating on the app, meaning less work, worse deliveries, and, ultimately, less money in his pocket. But solidarity, even from restaurant staff filling those orders, wasn't forthcoming. Delivery riders for apps like Uber Eats and DoorDash were the bottom of the food labor hierarchy even as, without them, countless restaurants wouldn't have made it through the pandemic and countless eaters would have gone hungry.

Out of this sentiment, shared by the city's 60,000-odd delivery workers, grew Los Deliveristas Unidos (Delivery Workers United), organized in 2021 in conjunction with the Workers Justice Project, an organizing hub for immigrant workers. Their demands, like those of Starbucks baristas and Waffle House grill jockeys, were for more dignity: access to bathrooms when picking up food, shelter from inclement weather while waiting for orders or eating, and better wages.

The last part runs headlong into delivery apps' business model. Apps often skirt worker protection and minimum wage laws by treating workers as independent contractors rather than employees. The app, their argument goes, is not a boss but an interface between clients and self-employed delivery riders. But given that the app dispatches riders, rates them, determines when they can log on and off, and controls their pay rate, this is a specious argument.

The apps also offload the capital costs of doing business onto the riders. Sosa tells us what we'd heard from other riders as well: To do the job, you need lots of gear, including a delivery-grade electric bike with at least two rechargeable batteries as well as cold- and wet-weather gear—an up-front investment that runs into the thousands. Recouping those costs means keeping a good rating on the app, which often means riding at breakneck speeds around the city trying to meet schedules that make sense only to a mapping algorithm that has never had to navigate through the chaos of traffic in the city. (As an aside, the actual number of delivery riders in the city is not known because there is no registry of independent contractors, but rather is estimated based on the number of registered delivery-grade e-bikes.)

After protests in front of city hall, New York City Council passed a resolution recognizing the delivery workers' demands, including setting a $19.56 hourly minimum wage. Then, in early 2025, DoorDash was forced to pay $16.75 million to compensate delivery riders for having used customers' tips to pay their wages rather than actually tip them. The lawsuit, brought by New York's attorney general, covered the years 2017 to 2019 and may have affected more than 60,000 delivery riders.[26]

Despite these wins, it hasn't been all smooth sailing since, an organizer with the Deliveristas told us, with the apps giving some riders fewer hours or even more onerous delivery schedules. Time for lunch must be booked in advance. But it's much better than it was before.

Another plus, says Sosa with a laugh, is that nobody calls them "essential workers" anymore.

The success in New York has not, however, been replicated at scale elsewhere. When California passed a law reclassifying independent contractors like Uber drivers and Uber Eats riders as employees, the app companies—transport network companies (TNCs) in policy jargon—engaged in a massive campaign to challenge the law via a ballot initiative on the 2020 ballot. The TNCs spent upward of $185 million on Proposition 22, which passed with a 59 percent majority, sending delivery workers back to their status as independent contractors. The National Employment Law Project, a labor nonprofit, likened the TNCs' tactics to those of the NRA (they meant the gun one, but the restaurant one would work, too).[27] Although some cities might see wins like those in New York, what might work best to break the TNCs' grip on the industry is not fighting for rights within the labor market the apps have created but in developing a worker-owned app that actually directly links consumers to riders, cutting out the Silicon Valley middlemen and their armies of lawyers and lobbyists.

OF ALL THE CORNERS OF THE FOOD SYSTEM, FARMS MAY WELL BE THE hardest workplaces to unionize. The reasons are complex and multifaceted, and the fact of it is perhaps surprising if you know your US history, which is, after all, littered with countless movements that organized agriculture. The Grange. The Farmers' Alliance. The Farm Bureau. The Farm-Labor Movement. The Southern Tenant Farmers Union. All of these made their marks on both history and agriculture.

Especially in the case of the Southern Tenant Farmers Union, many of those organizing were tenants, sharecroppers, subsistence farmers, poor, and without firm or permanent land tenure. But as the forces of competition, consolidation, and racial discrimination dispossessed many of those making up those movements, the few organizations that have survived—mostly the Farm Bureau—have tended to represent only the interests of the affluent commercial farmers to whom farmworkers sell their labor, the Earners we talked about in Chapter 2. Or, put differently in the context of this chapter: The best-organized group is the bosses.

Movements of farmworkers also have a deep if somewhat more obscure history that is tangled with the history of the demographic makeup of agricultural laborers. The Agricultural Workers Organization (AWO), for instance, was an integral part of the Industrial Workers of the World (IWW), or the Wobblies, as the radical international trade union was called, and it claimed as many as 100,000 members near the outset of World War I. Their numbers declined precipitously, along with the IWW's, in the next decade. The IWW was famously a multiracial union, but the ascendant trade unions of mid-century America were not. As a result, the racial and ethnic composition of the agricultural workforce—disproportionately Black and brown—ensured that farm-workers struggled without much solidarity from established unions.

Without that support, farmworkers had to forge their own unions largely on their own terms. They did this with the most frequency in the beating heart of American commercial agriculture, California. In the late 1960s, labor activists Cesar Chavez, Larry Itliong, and Dolores Huerta famously led Filipino and Latino farmworkers in California's table-grape industry on a nationally publicized strike. It resulted in a collective-bargaining agreement in 1970 that secured better wages and working conditions for around 10,000 farmworkers. Chavez and Huerta's United Farmworkers Association later merged with Itliong's Agricultural Workers Organizing Committee to become the United Farm Workers (UFW). Today, the UFW is a small part of the AFL-CIO, with only around 3,600 voting collective-bargaining members (5,600 if you count nonvoting retired members), but in 1975 its political muscle helped to pass California's landmark Agricultural Labor Relations Act (ALRA).[28]

The ALRA provides—at least in theory—labor rights for Califor-nia farmworkers. Those rights are not secured by the federal National Labor Relations Act of 1935, the New Deal Act that notably exempted domestic and agricultural workers from its protections. Even today, there is still no federal right to collective bargaining for farmworkers. In California, thanks to the ALRA, farmworkers *can* elect to form a union—indeed, it takes just two workers to petition an employer for better pay—and employers are prohibited from retaliating against unionizing workers or interfering with their elections. The act also requires that farms with more than 25 employees enter binding medi-ation to settle collective-bargaining disputes. All of this is overseen by California's Agricultural Labor Relations Board (ALRB). Passage of the law initially resulted in the UFW's membership surging to 60,000.[29]

But the law did all this *in theory*. The reality, as UFW's dwindling membership should indicate, turned out to be more complicated and disappointing. Democrat Jerry Brown signed the act creating the ALRB. His Republican successor, George Deukmejian, slashed funding to it. Overburdened and understaffed, the agency has since struggled to adequately monitor labor disputes in the state, especially as most farmworkers are now technically employed by third-party contractors, not directly by farmers, which leads to ever more baroque enforcement arrangements. Farmworkers, for their part, justifiably worry that the law's promise of freedom from employer retaliation is paper thin. A 2022 study by researchers at the University of California, Merced, found that more than a third of responding farmworkers would decline to file *any* complaint about their workplace's noncompliance with state labor laws, with the most common reason given being a fear of retaliation.[30] Making matters worse, in 2021 the US Supreme Court weakened the ARLA by striking down the law's guarantee that unions would have physical access to the farms where their members worked.

Those institutional problems dovetail with a complicating factor: California's farmworkers, like farmworkers everywhere in America, are overwhelmingly Latino immigrants. Fifty-nine percent of the respondents to the UC Merced study were not US citizens. Many do not speak fluent English. Many are undocumented. Those who are documented often have a short-term H-2a "Temporary Agricultural Worker" visa. They are a transient population that moves seasonally between farms of varying sizes all over the state, and up to Oregon and Washington, all of which makes organizing stable collective-bargaining units difficult. Added to all of this is a well-founded concern among immigrant farmworkers that retaliation will lead to deportation and family separation, a fear that 67 percent of respondents to the UC Merced study stated was "always" present.

The situation is even more daunting outside of California. Only twelve other states have followed California in extending collective-bargaining rights to farmworkers, and some states have other general labor protections that can apply to farmworkers, but these tend to suffer from the same enforcement problems that plague California. Other states have weak to nonexistent protections. And without regulation or enforcement, farmers have little difficulty intimidating workers who might complain to regulators (and firing those they can't intimidate).

This underscores the fundamental difference between farmers (landowners and bosses and only occasionally workers) and farmworkers (those who do the majority of the labor of producing food). Consider this example from the same Washington state asparagus industry that Paula DiPerna was writing about in the 1980s. Asparagus, like many crops, is highly seasonal, grows quickly, and becomes unusable if it is not promptly harvested. As a result, harvest is a labor-intensive sprint, with workers routinely putting in ten hours per day, sometimes seven days per week, all just to bring in the tender stalks that adorn your plate of steak. You might assume that any workers putting in more than 40 hours per week qualify for overtime pay under federal law. Alas, the Fair Labor Standards Act (FLSA) exempts farms from paying their workers overtime. To remedy this situation, in 2021 Washington state passed legislation to end the exception. The law—after a three-year phase-in period—would mean that workers previously making $1,400 a week during harvest season would stand to make more than $2,000.

Farmers recognized that paying overtime would cut deeply into their own profit margins. They claimed they would have to charge more for their asparagus to make up for the additional labor costs, which, they argued, would give competitors in places like Mexico a leg up. Therefore, many of them decided to sidestep the problem by hiring more workers for fewer hours, ensuring that few individual workers put in more than 40 hours, and they used labor contractors who often cut corners around minimum wage and minimum age regulations alike. That left all workers worse off. Following a similar rationale, farmers and the farm lobby tend to support immigration laws that maintain a supply of laborers willing to do hard agricultural labor at relatively low wages. Take Shay Myers, an Idaho asparagus farmer and self-described "staunch conservative" who penned a 2021 *Washington Post* op-ed calling for easier paths to citizenship for immigrant agricultural laborers for the simple reason that without foreign labor he cannot find "people willing to do the work."[31] He did not clarify how much he paid per hour.

One might argue that the Washington state law was flawed, but that's beside the point: The situation illustrates how and why the interests of farmers and their waged workforces are opposed. One group owns a business. The other group is employed by it.

That reality has forged some innovative approaches to collective bargaining in the absence of the formal legal right to do so. Boycotts and bad publicity have long been a major instrument for farmworker organizations, but none have wielded it more effectively than the Coalition of Immokalee Workers (CIW). Founded in 1991 by workers in Florida's tomato fields, the CIW is not technically a union, nor does it collectively bargain with growers on behalf of its members. It doesn't identify its members, in fact, nor does it have the same disclosure requirements that unions have. (Its critics, mostly agribusinesses and their lackeys, petulantly whine that the CIW is a de facto union if not a de jure one.) The CIW bypasses negotiation with growers, focusing its energies instead much further down the supply chain: all the way to the point of sale, where consumers touch the tomatoes. The CIW targets the biggest grocers and restaurants for unrelenting protests until the businesses agree to participate in the organization's Fair Food Program, an audited certification program for tomato growers which ensures that they provide their workers with a substantial wage premium and decent labor conditions.[32]

Many of the biggest names in food have signed on—some only after bruising protest campaigns, but others increasingly with the mere threat of protests. Signatories on the restaurant side include McDonald's, Taco

Bell, Burger King, Subway, and Chipotle. Grocers such as Fresh Market, Whole Foods, Trader Joe's, and even Walmart are part of it. So are food service giants Aramark and Sodexo. And since the launch of the program, cooperation between the CIW and those big buyers has resulted in $45 million in additional premiums to farmworkers above and beyond their normal wages. (The Fair Food Program requires growers to separately list the premium so that workers can see exactly how much money labor organizing is putting in their pockets.)[33]

This represents a monumental achievement, and one that is rightfully discussed with excitement in food media and writing. Mark Bittman's *Animal, Vegetable, Junk*, to take one example, names the CIW as an organization with great potential to transform the food system.[34] What's interesting about the CIW, however, is that it isn't advocating for a return to simpler, smaller, more local foodways of the type normally championed by Bittman. Instead, the CIW engages the world's largest tomato buyers—including the biggest of the big, Walmart—and leverages the outsize purchasing power of those buyers to favorably transform the labor conditions in Florida's tomato fields *at scale*. In fact, if you carefully examine the CIW's website, you will find no calls to source tomatoes from local farms or demands that Taco Bell sell less processed junk. The CIW isn't a foodie organization; it's about getting better pay and working conditions for the people who *work in industrial agriculture*.

Put differently, the CIW effectively exploits the consolidation of food production we described in Chapter 2 to get a better deal for its workers. Trust-busting to make those markets more competitive might reduce the CIW's leverage, which is why any enthusiasm for breaking The Barons, as Austin Frerick calls them, has to be accomplished along with expanding the *effective* rights of workers to organize, including, in agriculture, sectoral bargaining.

And that lands us on the bitterest irony of all and why sectoral bargaining, in particular, should be on the menu for farmworkers. Say the

United States enacted a law to enshrine the collective-bargaining rights of farmworkers tomorrow. Those small, local farmers you've read so much about—yes, the ones with the more labor-intensive, niche production methods who sell the good tomatoes at your local farmers' market—those are the workplaces that would be hardest to organize. Smaller workplaces are notoriously more difficult to organize than large ones. And farms tend to be more rural and geographically isolated than other workplaces, served by law enforcement that is more sympathetic to farmers, less to farmworkers, and not at all to labor organizers. The logistics of organizing thousands of workers dispersed across thousands of tiny workplaces spread across thousands of miles . . . "daunting" undersells it.

Improving pay and conditions for farmworkers will require ending the agricultural exceptionalism of American labor law and giving farmworkers the right to organize. It will also require investing in state agencies and building out regulatory capacity. That's what's needed to protect a workforce as complex, mobile, and dispersed as the one laboring on American farms.

AS IMPORTANT AS WAGES AND SCHEDULES AND WORKPLACE TREATMENT are to food system workers, one concern is always top of mind: safety. And for all the myriad dangers faced by food workers, from harvesters to forklifts to knives to wobbly ladders, increasingly one of the biggest threats to their safety is the environment itself. Most farmwork happens outside, and much of the most taxing work during harvest season happens in midsummer through midfall, often the hottest and driest period of the year. The sun, the heat, and the dust are the biggest dangers to workers out in the fields since dehydration and exposure can wear down toiling bodies. Research from 2015 suggests that farmworkers are thirty-five times more likely to die from heat than workers on any other job. It kills dozens every year. And an uncounted number more suffer from heat stress or just suffer throughout their shifts. As summers grow hotter, this problem will only get worse.[35]

Unfortunately, bargaining for better wages can be written into law or contracts, but the only reliable source of protection for job safety is strong legislation backed by strong enforcement. The federal Occupational Safety and Health Administration (OSHA), a branch of the Department of Labor, is the agency tasked with laying out rules governing workplace safety that employers must follow and with auditing and investigating workplaces for adherence to these rules. OSHA isn't perfect, but it's a crucial part of the regulatory state. The private sector simply cannot be trusted to ensure worker safety. Although many bosses grumble about OSHA, research has shown that its rules and inspections have reduced on-the-job injuries and deaths while causing no appreciable economic losses for companies. But OSHA was established to combat workplace dangers, not climate dangers, so it took decades of efforts by farmworker groups until rules were added to OSHA mandating protection from the heat. Signed into law by Joe Biden's administration in 2023, the rules are not yet in force. More alarmingly, OSHA is currently being targeted by the Trump administration's Department of Government Efficiency for budget cuts and firings that will diminish rather than expand its regulatory and enforcement capacity.

In the meantime, states including Florida and Texas, which experience stifling heat and have large outdoor workforces, have steadfastly refused to pass worker protections, bowing to the business lobby. These interests have trotted out the usual excuses about increased costs, which in turn would be bad for business and, in a cynical capitalist alchemy, therefore bad for the very workers being denied protection from the heat. The new federal law would override this in theory. But having rules and making sure the rules are respected are two different things.

We know this is the case because it's what's been happening in California for the better part of two decades. California's hundreds of thousands of farmworkers working grueling harvests in the fields—grapes in Napa, almonds in the Central Valley, and berries in Ventura County—have long grappled with heat risks. The Golden State, often an early adopter of progressive legislation—the anti-Florida, if you will—added heat protection to its state-level workplace safety agency, Cal/OSHA (a confusing acronym because it has nothing to do with the federal OSHA and is part of California's Department of Industrial Relations), all the way back in 2006.

The law requires employers to give workers mandatory breaks on hot days, including providing free cold water and a shaded spot like a tent. On days when the temperature passes the high nineties and heads into the hundreds, it's a commonsense solution that saves lives. Yet hundreds of farms fail to provide these protections. In 2023 alone, Cal/OSHA cited more than a thousand violations of the heat-protection code—and that's only the documented infractions. The agency is understaffed and underbudgeted, meaning that site visits and inspections cannot always cover all farms, especially in the more rural areas of the huge state. Farmers know this, so they often cut corners or flout the rules entirely; they and the middlemen they rely on to hire workers also often structure pay around a piece rate, where workers get paid per weight of produce harvested rather than by the hour, which disincentivizes them from taking their legally mandated breaks.

Like any other industry, agriculture needs binding worker protections and well-funded, robust enforcement against rule breakers, who in this case are farmers. Much US farm policy revolves around incentives for farmers to lean into good behavior. But on some issues we need to think not just about carrots but also about sticks. Implementing binding regulations and meaningful punishments—large enough financial ones that they could put rule breakers out of business—should serve to discourage noncompliance with laws. And we're not talking about forcing anyone into massive investments of money that would make business difficult: We're talking tents and coolers full of water. If farmers knew that not providing these basic, life-saving amenities could cost them the farm, they would provide them.

But the problem of heat, which will not go away on a warming planet, raises a more complicated question: Are there simply some jobs that should not exist? Is there work in food that is irredeemably bad, be it for workers or for society, and which we should think about phasing out or automating?

It's an uncomfortable question. Talk of doing away with jobs feels reactionary and may feel out of place in a consideration of the efforts to give workers a voice and fight for dignified labor. But how hot does it have to get for us to seriously wonder whether certain foods are worth the exposure of workers to chemicals, dust, and the sun? And what about jobs like those in slaughterhouses, regularly shown to be the most dangerous work in the food industry in terms of frequency of severe injuries? Even a unionized slaughterhouse will still involve awful work. And besides, given what we know about the food system, even achieving something like labor justice in an industry like meat does nothing to offset its environmental harms, never mind its harms to animals. Would we not, as a society, on balance be better off without slaughterhouses and slaughterhouse jobs?

Labor in and of itself is not a good thing. Good labor is a good thing. Dignified labor is a good thing. But perhaps not all types of labor can be dignified. And perhaps even dignified labor in awful industries should not exist because the harms of the industry outweigh the benefits of the labor to workers and the public. These are political questions with no easy answers, but after our years researching the food system, they are ones that we think at least deserve to be asked openly.

* * *

TWENTY-TWO MILLION AMERICANS LABOR IN THE FOOD SYSTEM. Their work is literally essential. The nature of their relationship to their employers keeps them underpaid and often underfed. But work that is currently bad doesn't have to be. Workers organizing for better wages and working conditions and unions want to turn their jobs into dignified ones. There is no single path for this to happen. Different industries and different types of workers face different challenges in different contexts.

One commonality among them is that their organizing ultimately needs to lead to legislation that enshrines their rights in the law. These are not fights that will be won in the supermarket aisle or food-science lab or academic research. They're fights that will need to be taken directly to the halls of power, where restaurant industry and agribusiness lobbyists hobnob with legislators. Passing those laws means pushing legislators not just to do the right thing but also to have the courage to stand up to powerful special interests with deep pockets. Time and again, these special interests, be they the farming, retail, restaurant, or app lobby, will claim that they cannot stay in business unless they give workers the bare minimum. Yet in other countries and in parts of the United States where labor has won victories, those same companies keep operating even as they provide better work environments. If there is a lesson from the food workers' movements, it's that making work political, and making big political demands, is the only way to make work better.

Child labor represents a hard limit to what guarantees of collective and sectoral bargaining for workers can accomplish. Child laborers don't need to be organized in their workplaces to better advocate for their interests; they need to be removed from those workplaces and put in school. At the end of the day, rules and a regulatory state with the capacity to enforce them are what will keep those tiniest of hands from harvesting the berries and asparagus.

This is the value of thinking about how institutions at scale shape the food system. Unions are institutions that protect and advance workers' interests and ensure that they are treated and paid fairly by their employers. Because they can also mobilize serious political power, unions have the ability to strengthen the various government institutions tasked with keeping food safe, air and water clean, and the kids in school and out of the meat grinders. The food bosses already know the value of scaled cooperation. That's why they form lobbying groups like the Farm Bureau and the National Restaurant Association. Those are the organizations cheering for DOGE to gut OSHA and for kids to stay in the fields. They know they're "stronger together," too, and that's why they fight to keep unions and regulators weak.

At the end of the day, then, we'd put this question to you: Who are *you* stronger with? The bosses? Or with people like Cindy from Waffle House and Christian the delivery rider—not just workers, but fellow citizens and fellow eaters?

Chapter 6

IN PRAISE OF
PROCESSED FOOD

IF THERE'S NOTHING MORE AMERICAN THAN APPLE PIE, THERE'S nothing more New England than pumpkin pie. It harkens back to a mostly mythical 1621 feast attended by hungry pilgrim colonists and their Wampanoag neighbors in Plymouth, just a jag south on the Massachusetts coast from what is now Boston. Whatever really happened that day, it may have been one of the first occasions when the Europeans sank their teeth into the sweet and mellow orange flesh of the New World gourd that Native peoples ate and cultivated widely throughout the continent. Then, as now, the feasters would have dined amid the beautiful colors of southern New England's changing leaves and the glorious light of the late fall sky. They probably had more seafood on the table than we do today, and they definitely had no football on the TV. But while pumpkins and New Englanders go way back, the pumpkin pie itself is a more recent arrival.

Widely circulated pumpkin-pie recipes date to the early nineteenth century, and regional chauvinism—Southerners preferred sweet potato pie—meant that it stayed a Yankee affair well into the twentieth century. The event that made pumpkin pie the national superstar that it is today was probably a marketing ploy by the canned-meat company Libby, McNeill & Libby. In 1929 it capitalized on the association of pumpkins with what the historian Cindy Ott calls "saccharine-sweet images of rural New England" by printing the now-legendary pumpkin-pie recipe on the side of a can of Libby's 100% Pure Pumpkin. It was an enormous hit, but the can doth protest too much. According to FDA regulations, the contents of a can of pumpkin can be made of a variety of squashes we don't conventionally call pumpkins, and, in fact, most canned pumpkin—and therefore most pumpkin pie—you've ever eaten is probably something called a Dickinson squash.[1]

But even knowing this, what's the alternative? Roasting and pureeing a pumpkin yourself? Feel free to whip up a crust from scratch, but under no circumstances should you use actual fresh pumpkin for the filling. Store-bought pumpkins don't have the right starch or water content for the custard. And don't think you can avoid this by buying a pie rather than making your own. Go to any bakery or grocery store in Boston the week before Thanksgiving and ask them what they use. Yes, even the fancy ones. They'll hit you with that New England honesty: It's from the can. No matter where you get it, if you want a pumpkin pie that tastes like proper pumpkin pie and summons thoughts of cinnamon, nutmeg, cloves, turkey-festooned sweaters, and family get-togethers, you will have to use the industrially produced goo.

Now here's the good news: The industrial goo is great. Libby's is now owned by the titanic Swiss food conglomerate Nestlé, which purchased the company in 1971. To this day, the only ingredient in its canned pumpkin is a mush of assorted squashes. They take millions of gourds grown on an astonishing scale; peel, seed, cook, and puree them; cram that puree into shelf-stable aluminum cans; and ship them off to grocery stores by the truckload. The stuff in those cans is nutritionally identical to a fresh pumpkin (ahem, squash) and will sit safely on your shelf for 900 days or more without spoiling. Meanwhile fresh pumpkins, as you will recall from your front step after Halloween, tend to rot after a week or so. And that pumpkin pie is a calorie- and vitamin-dense food, with most of its sugar and fat coming from the crust and the whipped cream. It might be a dessert, but it's also a solid serving of veggies.

Mainstream food writing and food media, including social media, has trained us to think of particular foods as good or bad, or even noble or sinful, and it derides industrial foods, like those that are processed and canned, as an unmitigated menace. But, like pumpkin pie, the truth is a bit more complicated. Just as we view the agricultural past through rose-tinted glasses, people often gloss over the health, safety, and nutrition problems that bedeviled our nation's food in the recent past. Industrial food production and processing and smart government regulation helped to eliminate those problems. Meanwhile, we over-complicate what constitutes a healthy diet, falling for fad diets, sup-plements, and knee-jerk condemnations of the unhealthiness of the American food supply and its reliance on "ultraprocessing."

But there is little secret about what a healthy diet should look like. And all the ingredients are already on supermarket shelves. Our ideal food system can be simplified to one core idea: *Everyone should have ready access to a varied diet of high-quality and nutritious foods all the time.* Never before have so many people had such reliable access to the foods they like to eat, in part because of industrialized processing that makes food safe, shelf-stable, affordable, and plentiful.

But by solving the problems of scarcity, the modern food system has created problems of abundance; diseases of poverty have for the most part been replaced by diseases of affluence. The food we eat now contributes to an epidemic of noncommunicable diseases, ranging from obesity to diabetes to heart problems. Forty percent of American adults are obese, 9 percent of the US population has been diagnosed with type 2 diabetes, and another 38 percent of the adult population is prediabetic. The direct health-care costs of these ailments run to around $600 billion annually.[2]

How can we preserve the tastiness, efficiency, affordability, safety, availability, and nutritional quality that industrial processing can provide without exposing eaters to harmful side effects? This chapter is about what the academic literature calls *utilization*, or how food that is available and accessible can become part of a nutritious, healthy diet. It explores the paradoxes of industrial-processing techniques that have, on balance, expanded the variety and quality of available food for most eaters while also making that food less perishable and more affordable. Yet those techniques have also sometimes created food that tends to be aggressively marketed, offers little nutritional value to consumers, and may pose some dangers to health. This is all complicated by rampant misinformation about healthy eating and fad diets. In this chapter we bust a number of nutritional myths and provide you with simple guidelines for healthier eating. But we also hope you'll take away a larger lesson: The industrial values of scale, reliability, and standards—values often reviled and dismissed in conversations about food—are necessary for an abundant food system, but they need to be kept in check by committed eaters and citizens so that they serve the common good.

IN A WORLD OVERFLOWING WITH FOOD ABUNDANCE, EATERS ARE FREQuently overwhelmed by both choices of foods and messages about what to eat. Thirty years ago, food historian Harvey Levenstein described this aspect of the modern food system as a "paradox of plenty" in his influential book by that name.[3] Twenty years ago, on the verge of the social media revolution, Michael Pollan, also in his book title, named it the "omnivore's dilemma." Both Levenstein and Pollan were concerned with how an abundance of dietary choices—what seemed to characterize the endless array of dining options in supermarkets and strip malls—can lead to bad dietary outcomes.

In the past two decades, the choices have continued to expand, but the rapid emergence of social media, accompanied by the fracturing of social consensus around expertise and authority, has only made things worse. If thirty and twenty years ago people were baffled or befuddled by dietary choices, they are now, to quote the Clash, "lost in the supermarket." In hopes of escaping dietary quagmires, many eaters embrace baroque and complex diets, often peddled by flimflam health influencers, that are hard to follow for very long and are not particularly healthy. When fad diets fail—as they almost invariably do—eaters find themselves right back where they started, only poorer for all the money that they spent on supplements and branded health foods.

We must grudgingly admit that it's hard to imagine writing a food book in 2025 that would prove as influential as foodie books in the 1990s and 2000s because of how social media, embedded advertisements, and paid health influencers now shape dietary advice and eating habits. To look out at the American food landscape is to be assailed by thousands of messages about food, from advertisements to claims made on packages to countless self-professed nutrition experts and gurus trying to sell you on a diet. Some of the best and simplest dietary advice, like national nutritional guidelines, winds up drowned out by websites, social media posts, blogs, podcasts, and books pushing complicated and often bad advice. When people are so exhausted by information saturation that they look to outsource the choosing to others, that's when the marketers, influencers, cranks, quacks, grifters, and con artists step into the breach and sell their fad diets, junk science, and nutritional supplements.

Our first bit of dietary advice in this chapter on what you should eat is this: Don't rely on social media posts from strangers, celebrities, and salesmen for your dietary advice.

Gwyneth Paltrow has told her millions of followers that her ideal lunch consists of bone broth, and she previously advocated a "detox" that excludes virtually every pleasurable food under the sun.[4] The detox idea itself, so popular in American food writing and marketing, is drawn from various riffs on "elimination diets." These were designed as medical interventions to identify and remove potential allergens from diets, but they have now become the basis for obscuritarian dietary regimes that randomly remove foods deemed to be "toxic" or "acidic" for no reason at all. One twenty-one-day program, promoted by Paltrow and other "wellness" influencers, tells eaters to avoid oranges, shellfish, soy and soy-derived products, almonds, and barbeque sauce. The avoidance of foods for alleged health reasons extends to grains, with a growing body of self-anointed fitness and diet experts, including the podcaster Joe Rogan, advocating for a virtually all-meat carnivore diet, ostensibly so that eaters get energy primarily from fats. It is worth mentioning that this meal plan cuts out the sugars derived from carbohydrates necessary for, among other things, proper cognitive function.[5]

The US weight loss and diet industry is a $90 billion per year business, but most of the information and diets it peddles are not based on solid, peer-reviewed evidence.[6] In fact, the peer-reviewed evidence we do have shows that virtually all "diets" fail to deliver better long-term health results than what you'll get by following the government's MyPlate plan or some similar basic nutritional guidelines. But what they lack in practicality, fad diets make up for with marketing and a storied and sordid history.

The British poet and politician Lord Byron famously managed his yo-yoing weight more than two centuries ago by an intermittent ascetic diet of biscuits and vinegar. It sometimes kept him trim, until he died at the age of thirty-six.[7] An early draft of today's low-carbohydrate diets was promoted by the British undertaker William Banting in his widely circulated *Letter on Corpulence* (1863), which cautioned against grains and bread.[8] Others argued for the opposite. In the United States, Reverend Sylvester Graham preached the benefits of "graham" (whole wheat) flour, benefits that he claimed were both digestive and moral. As food-studies scholar Kyla Wazana Tompkins notes, Graham maintained that a diet rich in minimally processed wheat was, along with hard beds and cold baths, a tool against regular "self-pollution." Later, Dr. John Harvey Kellogg made similar links between regular bowel movements and sexual continence, but his preferred diet food was cornflakes. Much like today, these diets, though contradicting one another in content, all promised eaters purity of food and soul.[9]

But many of today's diet gurus aren't just selling you on a meal plan. Many are also telling you not to listen to "the experts" while doing two seemingly contradictory things: directing you to eat only the pure and good foods that allegedly furnish complete health and nutrition while also selling you dietary supplements. But if the diet were *nutritionally complete*, why would you *need a nutritional supplement*? The answer is easy money and little legal responsibility.

The nutritional-supplement grift exists by dint of a—you may be noticing a pattern by this point—legal loophole. The Pure Food and Drug Act of 1906 appointed the USDA's Bureau of Chemistry to safeguard products sold as food or drugs within the United States, which in turn handed off that responsibility to the newly organized Food, Drug, and Insecticide Administration in 1927, which was eventually simplified to just the FDA and eventually moved to the Department of Health and Human Services. The FDA scrutinizes food to ensure that it isn't adulterated or toxic, and drugs are subject to rigorous clinical testing to demonstrate efficacy and safety. But supplements are neither food nor drugs as defined by law, so they are unregulated by the FDA—or anybody else. This exception was formalized in 1994 by the Dietary Supplement Health and Education Act, widely regarded as the outcome of extensive lobbying by the supplement industry.[10] Whatever claims to health benefits a supplement makes have not necessarily been confirmed by peer-reviewed trials or vetted by a regulatory body. We could chop up some grass from our lawns, grind it into a powder, package it into gelatin capsules, market it as Dr. Jan and Dr. Gabriel's Ivy League Smart™ Intelligence-Boosting Microgreens Superpill, and sell it at $49.99 for 25 pills. It would be a total lie and completely legal.[11]

Part of the morbid irony here is that dietary supplements are usually, to the extent they are even edible, highly processed foods. (Many, in fact, would qualify as what we discuss later in this chapter: much-vilified "ultraprocessed foods.") Thus, fad diets that promise freedom from the harms of industrial processing often sneak industrial processing back in—only it is industrial processing that is *less regulated* than the industrial processing the diets purport to avoid, which can make many supplements less pure, nutritious, and safe to consume than junk food.

This all leads us to some serious skepticism about health influencers as a class. Perhaps no one has taken the quest for healthy living to the extremes that tech entrepreneur and health influencer Bryan Johnson has. Having made hundreds of millions in various tech companies, Johnson has turned his attention to pursuing personal immortality through a baroque regimen of diet, exercise, supplements, and medical treatments, all of which are estimated to cost him around $2 million a year. Johnson follows a strict vegan diet of his own design to which he adds scores of pills of supplements. Oh, and he also regularly injects himself with his son's blood. You probably won't be able to access (or afford) the blood of the young, but much of Johnson's "Blue Print diet," including a bevy of supplements, is available for sale as branded products on his website.[12]

But what kind of good life is Johnson actually promising? Before a recent trip to India, he posted on social media that he planned to pack and eat his own Blue Print food during his six-day stay because the Indian food might disrupt his regimen. "Food is guilty until proven innocent," he explained.[13] Johnson will not become immortal. But worse, if you travel to India and miss the vada pav in Mumbai, the Seekh kebab in Delhi, a proper Goan vindaloo, and the hilsa swimming in Bengali mustard sauce all because you're eating out of your own suitcase, well, you may as well have never lived at all.

THE WELL-MEANING PURISTS ARE THE FLIP SIDE OF THE CHARLATANS and hucksters. They lean all the way into old-fashioned, unadulterated food, eliminating things from the common diet not with no good reason, like the grifters, but much more naively because these foods have been touched by the "industrial" food system.

So what should people eat in a food system where they can eat almost anything? The single best answer anyone we know of has come up with is Michael Pollan: "Eat food. Not too much. Mostly plants."[14] But Pollan's mantra, despite its apparent simplicity, needs some unpacking and updating.

The "mostly plants" part we cosign without reservation. And although it can be difficult in practice, we agree that it's good to avoid eating to excess. But the first part, seemingly the simplest, is complicated and cryptic. Pollan mostly meant that people should minimize processed foods in their diets, but things don't cease to be food just because they're processed, so it's here that things can veer into unhelpful sloganeering and confusion, including claims that average eaters should avoid all of the industrially produced foods, like packaged bread and peanut butter and frozen premade meals, which may be the cornerstone of their food habits or a necessity in a time-crunched schedule.

The turn toward ostensibly commonsense advice to abandon the very processed foods that make all diets, including healthy ones, affordable and palatable is ultimately as wrongheaded as that of the diet cranks. This is the realm of New Food Writers like Waters, Lappé, and Bittman and their many followers and imitators in food writing and on social media. Pollan explains that he differentiates between fresh whole foods and those that are in any way processed, and, further, between food and "foodlike substances." From this follows his second, even more often quoted commandment, which he meant quite literally: "Don't eat anything your great-great-grandmother wouldn't recognize as food."

Our great-great-grandmothers lived and died in Eastern Europe without electricity or running water (a few of Gabriel's did so in Ireland, where things weren't much better). They would not recognize a box of fusilli or, for that matter, a mango as food. They would have lived in a world on the cusp of many major breakthroughs in food storage, sales, and safety, but just shy of them. In this old world, there were no refrigerators in homes and no frozen foods, canning was a crap shoot, and there were no supermarkets, let alone an FDA to make sure your flour, butter, and milk were unadulterated.

Foodborne diseases like typhoid fever and botulism were common. Even more ubiquitous were diseases caused by malnutrition. There was endemic chlorosis, a form of anemia, caused by iron deficiency, that leads to chronic fatigue, a weak immune system, hair loss, and, in children, permanent physical and mental underdevelopment. Rickets, a debilitating childhood condition caused by vitamin D deficiency, results in bowed legs, stunted growth, and curved spines. Pellagra, caused by niacin deficiency, leads to severe rashes and sores and can cause dementia, mental underdevelopment, and death. Add to these goiters, the swelling of the thyroid gland caused by iodine deficiency; scurvy, caused by vitamin C deficiency; and beriberi, caused by thiamin deficiency. These are just the ones whose names you may recognize. Ever hear of marasmus? Be glad if you haven't.

Thanks, in part, to poor food, your great-great-grandmother's society was chronically unhealthy. In New York City in 1920, up to 75 percent of all children were to some degree affected by rickets.[15] Pellagra killed more than 100,000 Americans between 1900 and 1940, mostly in the rural South.[16] Average life expectancy in 1900 was 47 years.[17] When the United States initiated a draft during World War I, about 30 percent of men who were called up were deemed unfit for service, largely because of poor health attributable to dreadful nutrition.

The solution was a better diet offered by food processing.

By 1900, the widespread pasteurization of dairy products would have spared our great-great-grandmothers frequent salmonella, brucellosis, and listeria outbreaks from drinking raw milk. Pasteurization has saved countless millions of infants' and babies' lives, and it was probably responsible for reducing late nineteenth-century infant-mortality rates by over 25 percent.[18]

Then, in 1912, the Polish biochemist Kazimierz Funk published a paper on the causes of beriberi. Seeing that populations that ate brown rice rather than milled white rice had lower rates of the disease, he argued that a beneficial substance, thiamin (he called it "vitamine"), was removed during milling.

Funk's correct hypothesis that many diseases could be attributed to the presence or absence of isolatable substances in foods influenced early public health pioneer Joseph Goldberger, who studied pellagra in the American South. In contrast to the then-dominant view that pellagra was infectious, Goldberger believed it was caused by malnutrition, itself the product of poverty and unhealthy diets. The insight earned him four Nobel Prize nominations, but he died before biochemists managed to verify his claims when they isolated niacin.

Interventions to address these deficiencies were, in turn, made possible by the then still nascent field of food science, which found ways to isolate, produce, and then fortify foods, and especially staple foods, with vitamins and minerals. The lesson of the draft during World War I fresh on their minds, the federal government supported or even mandated some fortification. The results were stunning. Table salt with added iodine eliminated goiters almost overnight. Milk with added vitamin D dramatically reduced rates of rickets. Then, during World War II, the government mandated the enrichment of flour, making bread a daily source of iron, niacin, and other vitamins. Within a few years, pellagra and chlorosis were eradicated.

Preservation techniques like canning, which involve heating food enough to kill pathogens and vacuum sealing them to prevent recontamination, prolonged the shelf life of many foods, improving food access and reducing food waste from spoilage, especially once federal food-safety regulations came into effect after 1906.

These changes came alongside the emergence of a modern regulatory state, proving to many common people that federal oversight could tangibly improve their lives. However you feel about the regulatory state we've got, it's a huge improvement over not having one at all. It's only when you take for granted that your diet won't immediately kill you that you begin to ponder whether it might contribute to cancer or heart disease in a few decades. Synthetic preservatives may be to some extent unhealthy, but compare them to the common adulterants found in the "swill milk" of nineteenth-century New York City that sickened thousands of children: rotten eggs, chalk, and disease-ridden wastewater. Rotten eggs are arguably more natural than potassium sorbate, a common anti-mold preservative found in today's dairy products, but we would rather eat the potassium sorbate. A blanket dismissal of preservatives, additives, and processing obscures rather than clarifies the shared challenge of New York then and the United States now:

government capacity to keep truly harmful food out of supermarkets and fridges.[19]

Advances in transportation and refrigeration technology in the early twentieth century also gave birth to a radical new way to buy food—the supermarket—which also improved the quality and safety of food for most consumers. Selling a variety of foods that had previously been sold by specialized vendors spread around town, supermarkets delivered convenient access to fresh produce, dairy, eggs, meat, and canned and dry goods all under one roof. With an automobile, a person could shop for an entire week's worth of food for their family in a single trip, and with an icebox, fresh and perishable items might even last the week without spoiling. For many people, this expanded the kinds of fresh food they ate year-round while improved access to refrigeration and better food-safety standards in groceries reduced foodborne illnesses and spoilage.

Critics may assert that the pressure for standardized, widely available food has snuffed out the variety available to eaters and prioritized resistance to spoilage and visual appearance over things like taste and nutritional quality. This is the purported origin of the mealy and bland beef tomatoes, many of them produced in Florida, that one often finds in American supermarkets. Fair enough, but if you don't like that tomato, you can grab a pack of cherry tomatoes, a bag of romas, or maybe even a pricey heirloom. It's debatable whether the average tomato today is worse than the average tomato a century ago—this is often asserted, less often proved—but what's *not debatable* is that fresh tomatoes out of season were entirely absent from the American diet a century ago. The same goes for berries, broccoli, lettuces, cucumbers, pears, peas, apricots, and nearly all fruits and vegetables besides potatoes and onions. In fact, many of those fruits and vegetables were unavailable at *any time of the year* in climates that weren't suitable for their cultivation until the sprawling supermarket supply chains brought fresh food from distant locales. Today, most of these foods are available not just in their fresh variety but also as conventional ingredients in the various frozen, processed, and premade products that line the rows of the supermarket. And they all sit a stone's throw from a similarly diverse range of pastas, grains, rices, breads, and other staples.

Food historian Rachel Laudan heaps praise on the bounty of food made possible by industrialization and calls for an ethos of "culinary modernism" wherein we should embrace what she terms "high-quality industrial foods." We think our great-great-grannies would agree. They would think little of the people who turn their backs on modern food safety, such as the rising number of people flocking to raw milk. Raw milk, as they knew, but as some today have forgotten, is a reservoir of diseases, and in recent years has been linked to dozens of salmonella outbreaks that have sickened hundreds if not thousands of people. There are few things more irrational than turning your back on over a century of food-safety advances to purposely take risks with your and your kids' health in the name of eating more "natural" foods. If you tried singing the praises of unpasteurized milk to our great-great-grannies, they'd whack you with a wooden ladle for being a putz![20]

LAUDAN'S CALL FOR "HIGH-QUALITY INDUSTRIAL FOOD" DEFIES MUCH of the conventional wisdom about processing and large-scale thinking one finds in most food writing and foodie discourse. The ultimate stress test for culinary modernism would be if we could find exemplary qualities in even the most highly processed foods—those that Pollan famously decries as "foodlike substances." We set out to conduct just such a stress test—one not only of culinary modernism but also of Gabriel's willpower and sanity. Let us explain.

A common trope in food media is for a journalist to subject themselves to the diet they're writing about to document its effects.

Perhaps most famously, in the 2004 documentary *Super Size Me* Morgan Spurlock ate only McDonald's burgers, fries, and soda for thirty days to show the harms of junk food. He gorged on too many calories and, unsurprisingly, gained twenty-five pounds and spiked his cholesterol (though it was later revealed that Spurlock was also dealing with a variety of other compounding health issues that may have been to blame).[21] More to the point, Chris van Tulleken, the author of the bestseller *Ultra-Processed People*, embraced a diet of 80 percent ultraprocessed food. His rations of sugary breakfast cereals, ready-to-eat English breakfasts, precooked lasagnas, and up to six Diet Cokes a day—equally unsurprisingly—led to a series of health ills, including weight gain and constipation that led him to abandon the experiment before thirty days were up.[22]

Van Tulleken's book focused on what is increasingly blamed as a major contributor to collective poor dietary habits and resulting poor health: ultraprocessed foods (UPFs). It's a term you'll find splashed across headlines and book titles, meant to communicate the dangers that industrial-processing techniques pose to human health and nutrition. UPFs do have a formal technical definition we'll get to later in this chapter, but in the media the term is often a sloppy stand-in for "bad" or "unhealthy," attached to everything from plant-based meat alternatives to packaged sliced bread to cooking oil. But is it really all bad?

One of the problems with stunts like van Tulleken's is that a sample size of one is statistically meaningless. It's storytelling and not science. But we *are* suckers for a good story. So we indulged in some of our own food autoethnography: Gabriel would spend a month eating only UPFs. None of this weak-sauce 80 percent business. But rather than eating processed meat and sweets and microwave dinners, Gabriel would eat only products made by Huel, a UK- and Brooklyn-based company that makes nutritionally complete vegan meals and meal replacements. Its sales pitch is simple and obviously aimed at young professionals looking to Marie Kondo their food choices: "Swap out lunch, swap in Huel."[23]

Every serving of Huel has the same calorie count (400 calories) and a similar nutritional profile, designed to optimize the delivery of both macro- and micronutrients. It comes in three formats: large, nearly indistinguishable shiny resealable black plastic bags of meals and protein powders and preportioned microwavable meals in one-use paper cups. Scoop, add water, then shake and glug or microwave and chow. Or just get their ready-to-drink protein shakes and chug.

Gabriel usually spends quite a bit of time cooking and eating, and with Huel, he cut it down to thirty minutes over the course of the day. Most Huel meals run somewhere around $4 per portion, which means that Gabriel's entire monthly food budget was just under $800—almost exactly the $779 that the average American spends on food every month between groceries and eating out.

Here's what Gabriel was like before the diet. His iPhone tells us he averaged 6 miles' worth of movement every day, roughly split between running for exercise and walking around town and campus. He's kind of obsessed with kettlebell weight training. He packs about 190 pounds onto a six-foot-one frame and has a resting heart rate under 50 BPM and blood pressure around 100/60—not bad for a guy in his mid-forties. He eats a mostly plant-based diet as well as small amounts of eggs, dairy, and fish. As we pondered his all-Huel future, we wondered: Would nutritional restrictions undercut his fitness? Would he meet his protein needs? Would he feel tired or weak? Or nauseous? Would he gain weight like other UPF self-experimenters?

The rules of the experiment were simple. For an entire month, only Huel passed Gabriel's lips save for sugar-free breath mints, toothpaste, salt and pepper, water, coffee, and Diet Coke. That's about a 99 percent UPF diet. Meanwhile, Gabriel maintained both his regular caloric intake and level of physical activity. He checked his vitals and blood before and after the experiment.

Let's get straight to the downside: The all-Huel diet quickly became an excruciatingly unpleasant eating experience. You should under no circumstances attempt to replicate it. This might sound obvious, but in the early goings, it could be kind of fun. Gabriel approached these early meals with plucky, exploratory cheer. The (faux) Chick'n & Mushroom captured the umami richness of regular chicken and mushroom, just as the also excellent Pasta Bolognese nailed the flavor of the red wine that imparts a normal Bolognese with its distinctive richness. The Banana Pudding shake was smooth and mellow.

But by the second week, the inherent limitations of the menu became apparent. Gabriel could deal with monotony. But what he couldn't deal with were the peas, which are the primary ingredient in most Huel meals. Pea protein, in particular. It's flavored and textured differently, and it's supplemented with other ingredients, but at the end of the day, a Huel diet is essentially a pea diet. Once you become aware of this fact, it's impossible to forget it. The promising Strawberry Shortcake shake became a strawberry-flavored pea shake. The Mexican Chili became Tex-Mex peas.

Over the month, the experience of eating nothing but various incarnations of peas was maddening. Meals, usually something to look forward to, were increasingly a source of dread.

While Gabriel's Huel ordeal hardly proves anything, it does prompt us to raise some points that the rest of the chapter will examine in more detail.

First and foremost, don't ever underestimate the hedonic aspects of eating. If people don't enjoy the food they're eating, they probably won't eat it for long. Gabriel pulled it off for a month, but even that was tough.

But second, and more importantly, it didn't appear that these ultra-processed foods were inherently unhealthy. The all-Huel diet seemed perfectly fine. Gabriel started to lose weight while maintaining his work and workout schedules. His blood work and vitals showed no change at all.

Much like how it matters more to the environment what is grown than how it is grown, we would argue that for nutrition it matters more *what is processed* than *how much it is processed*. In the fall of 2025, *Consumer Reports* tested protein supplements and warned of high levels of lead in many of them, including Huel's Black Edition, but this was based on an unrealistically strict threshold for lead, well below what the FDA recommends as a limit for women of child-bearing age. This confusion is a prime example of the need to regulate supplements as stringently as food.[24] So while we cannot recommend an all-Huel diet, we would note that neither does Huel. One could easily imagine a pleasant, satisfying, balanced, and healthy diet that incorporated Huel as a component, alongside other processed and fresh ingredients as well as odd treats. In fact, despite the odiousness of the month, Gabriel still eats various Huel products, including that banana pudding protein shake after workouts.

GABRIEL'S HUEL EXPERIMENT WAS FUN (FOR JAN TO OBSERVE), BUT it also raised some questions about what, exactly, sound culinary advice would be that works for most people. "Eat Huel" might seem to push back against "Don't eat UPFs," but it's not the general dietary advice most people need. For more comprehensive and peer-reviewed wisdom, we decided to consult nutritional experts at Tufts University's world-renowned Friedman School of Nutrition Science and Policy.

What those experts told us was the polar opposite of what most health influencers and culinary naturalists argue: A good, affordable diet for most consumers entails taking advantage of high-quality industrial foods, not shunning them. In fact, a tendency to shun high-quality industrial foods—and consequently making your diet narrow, rigid, and impractical—may be one reason that so many dieters belly flop.

Cultivating balance, flexibility, and moderation is often more important for long-run health and sustainable dietary change than finding that "one weird trick" promised by adherents of any of a slew of fad diets.

The Friedman School of Nutrition Science and Policy is part of the Tufts Medical Center in the beating heart of Boston, and it occupies a towering modern building to the south of Boston's Chinatown and the Boston Common. Its professors have on numerous occasions served as members of the Dietary Guidelines Advisory Committee, a panel of experts that shapes the federal government's attempts to map out a healthy diet for Americans.

The USDA has published dietary guidelines since 1943. For decades, those guidelines were represented by a pyramid that arranged food groups in an ascending order of how many servings you should eat and how often you should eat them. Grains at the base, followed by fruit and vegetables, dairy, and protein, with fats, oils, and food high in sugar at the pyramid's apex. It was mostly accurate, but it did visually give the worst foods pride of place at the peak, and the focus on daily quantities instead of individual meals made it due for a revamp. In 2011 the pyramid was replaced by MyPlate, which represents the dietary guidelines as a plate with the food groups laid out on it: half of the plate dedicated to veggies and fruits, a third to grains, and the rest to protein, accompanied by a glass of milk or fortified plant-based alternative. The USDA suggests that this general distribution should prevail in most meals.

The USDA's process for arriving at its guidelines is not ideal. Nutritionists play a role, but so do industry groups and the elected officials they lobby. The animal-agriculture industry successfully fought to have dairy included as its own separate food group. Meanwhile, the scheme confusingly shoved protein-rich whole grains and legumes into other food groups, leaving the "protein" food group to be deceptively composed of mostly meat and fish. Those are problems worth addressing in future iterations. But as a broad-strokes heuristic, MyPlate works fine.

There is no one-size-fits-all dietary size—that depends on the eater's age, physical size, level of exertion, and other factors—and MyPlate recommendations can be adapted to a range of ages and calorie-intake levels. But for the sake of our example, let's use a 2,200-calorie diet which, following MyPlate, over a day's eating adds up to two cups of fruit, three cups of veggies, seven ounces of grains, and six ounces of protein (plus three servings of dairy or fortified soy or other plant-based alternative).

Consider these three sets of meals as examples of what that could look like.

A peanut butter and jam sandwich on whole-wheat sliced bread, a cup of coffee with a generous pour of skim milk, and an apple for breakfast. A banana and a granola bar for a snack. An egg-salad sandwich with lettuce and tomato and some carrots or salted almonds or even a small bag of chips for lunch. A latte to get you through the afternoon. A big plate of spaghetti Bolognese and a glass of red wine for dinner. Drink plenty of water.

Chilaquiles with tomatillo sauce, beans, onion, cilantro, cheese, and cream for breakfast. Some precut papaya for a snack. A couple of tacos with the fixings from the taco truck for lunch. Make sure to get the guacamole and some *cebollitas preparadas* on that plate. Rice, beans, and fried plantains with your choice of hot sauce for dinner.

Overnight oatmeal made with oat milk, chia seeds, mushed banana, and blueberries. Green tea. A vegan protein bar for snack. A big salad for lunch. Tofu steaks, mashed sweet potato, and grilled asparagus for dinner. Kombucha.

You'll notice that the MyPlate guidelines can accommodate an endless variety of meal plans, with the specifics fitted to the preferences and cravings and cultures of eaters and their families. As Dr. Suzi Gerber of Tufts reminds us, a healthy diet should "serve pleasure as much as health," especially if people are to maintain long-term nutritious eating habits. Gerber is a USDA-funded researcher at the Friedman

School and something of a food polymath. Her interest in the role that pleasure plays in good nutritional outcomes stems from the fact that she worked as a professional chef before she was a nutritional scientist. Today, she publishes peer-reviewed scientific papers, engages in food-policy advocacy, writes cookbooks, and runs a popular Instagram account dedicated to plant-based eating. Gerber thinks that health and nutrition discourse often unhelpfully conflates healthy eating with ascetic denial. By contrast, she maintains that a healthy diet requires balance and variety that regularly integrates flavorful faves and leaves room for treats. However, Gerber's research shows that people do often lack the support they need to stick with the dietary habits to which they aspire, not only in terms of choosing and cooking food but also in terms of giving their bodies and taste buds time to adjust to healthful changes. Much of Gerber's policy research focuses on how governments and large institutions can better furnish eaters with the support they need to stick with the healthier diets they want. And healthier diets are definitely something most Americans need.

The best tool to judge the healthiness of diet at a population level is called the Healthy Eating Index (HEI). Based on a large survey of American eating habits, the HEI compares Americans' diets to the nutrition recommendations of MyPlate. The results aren't great. The top score in HEI is 100, with anything over 80 rated as a healthy, nutritious diet (an A grade). The mean HEI score for Americans between the ages of nineteen and fifty-nine is 57, which would translate to a D grade.[25] The results show that many Americans eat too few fruits and whole grains and too many items packed with saturated fat and sugar. Surprised? Probably not. But you may be surprised by *why* it's the case.

One common theory is that healthy food is more expensive than unhealthy food. Accordingly, budget-conscious munchers opt for calorically dense but nutritionally deficient foods that are dripping in fat and sugar, like *Homo economicus* at an all-you-can-eat buffet.

This common theory is mostly wrong. As Gerber notes, in addition to falsely assuming that healthy diets are pleasureless miseries, people assume they're more expensive. There *is* a link between poor nutritional outcomes and food insecurity, as we discussed in a previous chapter. But most people aren't food insecure, and once you remove the food insecure from the pool, the data tell a different story. HEI surveys show that income has only a minor impact on the healthiness of diets. The discrepancy between the richest Americans and the poorest is only 5 points. In other words, all Americans eat relatively poorly. And rich Americans, who could in theory eat the healthiest diets imaginable, for the most part don't; if the average American scored a D, the richest Americans still managed only a C–. (Notably, although wealthier Americans don't eat significantly healthier diets, they do have better health outcomes for reasons unrelated to diet, such as less exposure to toxic pollutants, lower levels of stress, and better health-care access.)

If the costs of healthy diets aren't the driving factor of our poor eating, what is? We sat down with Friedman School economist William Masters to find out.

"A healthy diet is actually pretty simple," Masters tells us. "The average American can access a healthy diet more easily than almost anyone ever before in history. The problem is the displacement of healthy diets by unhealthy ones." The biggest culprits are the ubiquitous advertising for and the presence of unhealthy foods in food landscapes. That pervasiveness shapes habits and taste buds, prompting people to eat unhealthful diets or overeat without even thinking about it.

Think about your most recent trip to a supermarket. You will recall you were bombarded with an astonishing array of products in eye-catching packaging that promised pleasure and health. Although many of these products are tasty, their health boasts can be specious and deceptive. The $0.99 can of healthy black beans looks downright listless next to the vibrant box of Fruit Loops that cheerfully claims to be "a good source of 9 vitamins and minerals." Sodas, sugary cereals, chips, desserts, and countless other indulgences choke out options that are both healthier and more affordable. That's also the case in many restaurants, where sugary, fatty, fried options predominate. "A healthy diet is absolutely within reach," Masters tells us. "But it's swatted away by the utterly compelling one."

Many modern food landscapes habituate us to patterns of eating that make exceptional treats into routine foods. It's not necessarily the price tag of the treats at the grocery but how little friction we experience grabbing them. It's the difference between having a to-die-for triple-fudge Belgian chocolate brownie fondant dessert at a favorite restaurant one weekend a month and loading up on one-bite brownies every time you wander past the plastic tub on the kitchen counter. It's the difference between walking to a corner store to buy a can of Coke when you crave it and stocking a fridge in your garage with dozens of them. It's the fundamental conundrum at the heart of the paradox of plenty: Many food environments provide broad access to affordable healthy options, but those same food environments also steer eaters away from those options.

There are policy interventions that could improve (and have improved) food landscapes. For instance, once there was enough proof of the noxious and long-term health impacts of trans fats, the United States first moved to clearly label them and then banned them outright. There have also been many local efforts at the city and state levels to tax sugary drinks. All of these policies have merit, but at the end of the day, in a market economy we're unlikely to ban all foods that contribute to an unhealthy diet, which means that, like it or not, eaters will have to make good dietary choices if they want to avoid bad outcomes.

To shop healthfully, Masters suggests disregarding the health boasts found on packaging, which are mostly bunk, and just using the USDA's MyPlate app to achieve nutritional balance without overeating. He recommends that you "just buy the cheapest, most generic, simple things that are in each food group. It doesn't have to be expensive, but it does require you to walk past the aisles of relentless advertising, drive past all the billboards for fast food and premium items, and tune them out." Don't deny yourself "the occasional ice cream cone or other treats," but do treat them as exceptional. Following even those simple guidelines "can save [you] a lot of money and you can be healthy," Masters concludes.

None of this will instantaneously sideline drivers of unhealthy eating that are embedded, quite literally in the case of supermarkets, in the structure of everyday life, which returns us yet again to the thorny problem of personal agency. As we noted about meat consumption, it's an error to frame the problem in terms of individual choices versus impersonal systems and structures. Food shoppers have agency, and they need to use it to make healthier choices, all while we should strive to create food environments that make exercising that agency easier. In discussions of nutrition and obesity, this error typically pits people who say that individuals need more discipline against those that say personal choice isn't up to the task and people need help. Don't fight; both are right!

Consider the case of a promising medical solution to chronic obesity and type 2 diabetes, a class of drugs called GLP-1 agonists, such as Wegovy and Ozempic. They work by injecting a "glucose like peptide" into the bloodstream that moderates blood-sugar levels and, essentially, tricks your brain into feeling sated. That, in turn, weakens the pull of impulsive and risky eating. In short, people desire to eat less, so they do, and that causes them to lose weight.

Numerous peer-reviewed studies confirm that GLP-1 agonists are effective at treating obesity and do not appear to have major health downsides: Patients who took them exhibited sustainable weight loss as well as related improvements in health outcomes that outweighed any tangible biomedical risks the drugs carried.[26]

If you think that this rebuts the most commonsense medical prescription for obesity—eat less, exercise more—you're looking at it wrong. GLP-1 agonists aren't a replacement for the prescription that your gym teacher, doctor, and granny would all tell you is basic common sense; they are a biomedical tool that helps patients follow that prescription. To put it differently, eaters need discipline to avoid overeating and stay active, and GLP-1 agonists are drugs that can give patients that discipline.

There are still unanswered questions about GLP-1 agonists. People may object that dampening desire drains life of its meaning and drains eating of joy. Others dislike the dependency on global pharmaceutical firms that wide use of the drugs could create. Beyond these abstract questions, there are complicated policy and medical ones: Are the drugs being prescribed appropriately given clinical results? Should the federal government increase access to the drugs to improve population-level health outcomes? How? And who will pay for it?[27]

Definitive answers to these questions are probably premature, but they do help us understand something more basic about diets and effective dietary change. GLP-1s make it easier for people to restrict caloric consumption and follow the generally good advice of "eat less, exercise more." Similarly, when restrictive diets do help people to lose weight, it is often because of all the foods the diets exclude, not the handful of foods the diets prescribe, because the restrictions make it practically harder for people to eat as much as they would otherwise. Remember how Gabriel started to lose weight on his all-Huel diet? That was because its proportioned meals made calorie tracking easy and he simply wasn't allowed to eat other foods he encountered throughout the day. So he just ate less and didn't snack. Following any draconian diet, regardless of its contents, makes it harder for people to find food, and it effectively limits how many calories they tend to consume. If you can stick to it, it works; but the weirder the diet, the harder it is to stick to it in the long term.

That's largely why, beyond all the theatrics and con artistry, we don't think that weird diets are good for most people, a view shared by all of the experts at Tufts with whom we spoke. Most consumers would do well to follow Masters's shopping advice and keep it simple, affordable, and generic, all while keeping it varied. Eat mostly the foods that you (and your family) enjoy in the quantities that MyPlate—or similar national dietary standard or something like the EAT-Lancet diet—recommends. Don't be draconian or self-punishing about it. And pay attention to (and perhaps seek to avoid) food environments where you tend to overindulge. One of the biggest studies of healthy aging, covering more than one hundred thousand people and published in 2025 in *Nature Medicine*, more or less confirmed this, restating what should be common sense: "Higher intakes of fruits, whole grains, vegetables, added unsaturated fats, nuts, legumes and low-fat dairy were associated with greater odds of healthy aging."[28] Eating healthy may not be easy for most people, but it doesn't need to be complicated or expensive.

ONE RECURRING THING YOU'LL NOTICE ABOUT BOTH THE FAD-DIET gurus and the culinary naturalists is that they tend to present questions about diets not as *analytic puzzles* but as *moral tests*. A "good diet," for many of these diet prescribers, is ultimately about having the personal discipline to obtain dietary purity in a contaminated world. For us, as for Masters and Gerber, it's an empirical question. However, the confusion around the normative and the analytic is a broader problem for many food writers and journalists, and it finds no greater illustration than in confused and confounding writing about "ultraprocessed foods" (UPFs).

In 2009 a group of nutritionists led by Professor Carlos Monteiro at the University of São Paulo in Brazil created a classification system to parse foods into epidemiological categories suitable for government public-health and food policies. It's called Nova (even though it's often capitalized as NOVA, it's not an acronym but simply the Portuguese word for "new," as in "new classification"). Monteiro's team was aiming for categories that could describe the extent to which a food stuff was processed for the sake of crafting food policies—such as linking broad consumption patterns to obesity in particular populations or geographical areas—and not to measure any particular food item's nutritional quality or implications for human health.[29] This point, though clear from the team's initial research design, is now often lost in public debates and academic conversations alike.

Nova divides foods into four categories in ascending order of processing. The first group contains unprocessed or "minimally processed foods." This includes fresh and frozen fruit; raw grains; and fresh meat, eggs, and milk. The second group has processed culinary ingredients that are ground, dried, or otherwise transformed from an inedible original form into something you can eat, like unenriched flours, seed and vegetable oils, and salt and sugar. The third category is where you find processed foods that combine ingredients from categories one and two and are subjected to further techniques such as baking, canning, or boiling before sale. This is often where foods are made shelf-stable. Bread, pasta, cheese, and canned pumpkin and tuna are here. Most home-cooked meals combine ingredients from these categories.

This brings us to category four: "ultraprocessed food." The full definition is very long, but in short it describes UPFs as those that include further processing and ingredients that are not found in most kitchens; that use novel stabilizers and emulsifiers; and that extract, repurpose, and fraction foods into particular nutrients like whey or soy-protein isolate.

Scientific inquiry and public health policy both need some degree of schematization and simplification, and Nova provides it for levels of processing, which can also aid in some forms of simple public communication. You might call it Monteiro's Razor: If you find it in the produce section of the grocery store or if it's a fresh-baked loaf, it's probably not a UPF; if it's canned or has a long list of ingredients, it probably is one.

But beyond this level of generality, Nova is not a razor but a potato masher; it takes many diverse foods with variable nutritional qualities that are processed in distinct ways and then smushes them into a mush. Nova's infamous fourth category contains items as disparate as Impossible burgers, Oreo cookies, baby formula, and Coca-Cola. Each of these has vastly different nutritional profiles and uses, and is made using distinct ingredients and processing techniques. An Oreo is mostly sugar, and we'd wager that whether you eat an Oreo or the equivalent four teaspoons of sugar (yup), it will be just as bad for you. And baby formula is a crucial part of many newborns' healthy dietary needs, but giving your newborn Coca-Cola is child abuse. An Impossible burger might have high levels of sodium, but it's a tasty source of protein.

Now, a number of widely publicized studies have purported to show the negative health effects of UPFs as a category. But do they? Let's examine three of the most robust ones.

The first was published in the prestigious *British Medical Journal* and used data from a large cohort of Americans to assess the links between diets and health outcomes.[30] The widely reported top-line finding was that those participants who ate a diet high in ultraprocessed foods had a slightly higher mortality rate than those who didn't. Scary! But dig a little deeper, and, in fact, the correlation between ultraprocessing and mortality was statistically significant only for processed meat and seafood as well as sugary drinks, cereals, and desserts. All other UPFs made no statistically significant contribution. The study merely confirmed what we have known for decades: Too much meat—especially red and cured meats—and sugar are bad for your health.

The second study, led by members of the original Nova team, was published in *The Lancet Regional Health*. It used data from a large cohort of Britons to examine how plant-sourced UPFs affected cardiovascular health.[31] Again, the top-line finding was that high UPF intake, even from plants, was correlated with bad heart health. So Impossible burgers are bad, right? Not quite. The plant-based UPF category included shelf-stable bread, pastries, fries, soft drinks, and distilled alcohol. Plant-based meat alternatives made up only 0.2 percent of participants' diets! Counting Coca-Cola, Oreos, and vodka as plant-based foods is, while technically accurate, likely to be misinterpreted. And, indeed, a study of soy-based meat alternatives published in 2022 in *Advances in Nutrition* found they were perfectly heart-healthy and definitely more so than many unprocessed (Nova 1) red meats, and a 2025 article in *Frontiers in Nutrition* argued that plant-based meat alternatives are a healthy source of protein.[32]

There is, however, one very important study that gives hints about *why* and *which* UPFs can be bad for our health. In an experiment, nutritionist Kevin Hall gave two groups of people two diets, with the dieters responsible for moderating food intake themselves. The diets were nutritionally identical, but one had no UPFs, and the other was 80 percent UPFs. The UPF-gobbling group gained weight. Why? The people eating the UPFs tended to eat 500 calories more per day than the other group. The UPFs, designed to be "hyperpalatable," bypassed people's satiety instinct.[33] These foods tended to be easy to chew and swallow or addictively crispy and crunchy, meaning that they went down too quickly or in higher quantities than people might otherwise have eaten them.

And it's here that we get to the problem with UPFs: It's mostly not the "ultraprocessing" itself that's the problem as much as the fact that ultraprocessed foods can be extremely tasty, easy-to-eat, convenient vehicles for unhealthy ingredients. One recent study suggested that UPFs high in carbs and fat could even become physically addictive.[34]

To understand the potential long-term impacts of this, we sat down with Fang Fang Zheng, an epidemiologist at the Friedman School.

Zheng's research focuses on the long-term links between diets and health, and specifically with diseases like cancer. The top line of her findings is that the biggest contributors to diet-related disease risk, including cancer risk, are red and processed meat and sugar-sweetened beverages like soda, coupled with low intake of whole grains and fruit and veg: the same culprits identified by the HEI index. Where UPFs exacerbate this situation is that many contain exactly these ingredients, which they pollinate throughout our food environment. This is especially a problem for children and young people, whose diets are now composed of 67 percent UPF, mostly from premade meals and sugary and salty snacks. Yet she is wary of relying entirely on Nova: A recent research study she coauthored shows that consuming ultraprocessed fruits and vegetables *decreases* long-term health risks. If anything, these findings support MyPlate's emphasis on the quantities of types of food, not the levels of processing. Her solution is a commonsense epidemiological schema that marries the original intent of Nova (a taxonomy of the level of processing) with the actual nutritional properties of foods, which would basically mean that a net cast to identify Oreos and Fruit Loops and microwave pizza wouldn't also scoop up Impossible burgers or microwavable saag paneer with brown rice as bycatch.

All things being equal, then, we think a precautionary principle suggests moderating UPF intake, especially when those UPFs are processed meats, high in sugar or salt, and fit the pattern of "hyperpalatability." In other words, when they're what we used to call junk food, and which you can readily identify without wondering about the unseen level of processing or ingredient list of any given food. But that's about it. Eliminating UPFs (or junk food for that matter) completely from your diet will no more guarantee health than eating them in light moderation will kill you. Don't let the fearmongering about "bad foods" get in the way of following basic advice about a healthy diet or make you second-guess your fortified packaged bread, peanut butter, a frozen bean burrito, or a well-formulated meal replacement like Huel.

Much like the term "GMO," "UPF" is a concept that can clarify the diverse processes and technologies that shape what lands on your plate. But be wary of slipping too quickly from the descriptive and analytic to the prescriptive and normative. When it is used in a narrow technical sense, the concept of "ultraprocessing" and the scientific framework behind it can be valuable even if it's likely far too simplistic. But because it's often misused and it sounds scary, it leads people to draw false inferences about the nutritional and health consequences of their diets all while promoting moralistic polemics. The Non-GMO Project, the group that labels products as being free from GMOs, thereby likely doing more to create anti-scientific stigma than promote anyone's health, has now promised to move into labeling products as UPF-free. This again would likely do more harm than good, and it represents a misguided reductionism in thinking about food, obscuring actual nutritional content and safety behind fears of an acronym.

SPEAKING OF LABELING, THIS IS ONE PLACE WHERE POLICYMAKERS could implement an easy change that would assist consumers: require food producers to put the nutritional labeling on the front of the package and to make the labeling easier to read. Food companies may complain that forcing them to foreground unflattering nutritional information will hurt their bottom lines and encroach on their freedoms. We think that's ludicrous. Companies already foreground what they see as beneficial nutritional qualities of products; it's only fair they do so with *all* nutritional qualities. If nutritional labeling is worth doing, it's worth doing in a way that enhances, rather than undercuts, the communication of nutritional information. We might even go one step further and take a page from the Mexican government, which requires products that have an excess of sugar, fat, salt, and calories to put a small seal to this effect on their packaging. American companies might object that they'd never do this, but they already comply in Mexico.

A second policy intervention that would make a big difference would be bringing supplements under the regulatory oversight of the FDA, no matter how much the "wellness" lobby would fight against it. Quite simply, supplements' health claims would have to be vetted by the government, subjected to the same level of scrutiny as novel ingredients. If they fail, too bad. This would have the double benefit of saving gullible consumers lots of money every year and cutting into the profits of some of the worst peddlers of food disinformation out there. But yes, to the extent that more study and regulation is needed, the FDA needs to be staffed fully and effectively.

Alas, the opposite is happening. Recently, the Trump administration announced that it would be laying off 10,000 workers from Health and Human Services, including 3,500 from the FDA. Although none of those 3,500 are currently employed as food reviewers, targeted employees work in positions that provide reviewers with the support they need to do their jobs effectively. Those cuts should be overturned, whether under this administration or the next. Having a regulatory state that can detect and remove dangerous food from menus and tabletops is a prerequisite to it doing so.[35]

We're not giving all industrially processed food a free pass. There are scientists and writers out there who make bank trying to discredit all critiques of food processing. Food scholars like NYU's Marion Nestle argue that conflicts of interest are prevalent in nutrition science, with many studies and scholars funded by the food industry.[36] Often, those folks want to defend food companies' bottom line, and those food companies don't want you asking questions about what you're eating. That's not what we want. We want you to make informed decisions, and we want policy to make it easier for you to do so (to say nothing of keeping harmful food out of the food system entirely). But to do that, we advise getting away from using "processing" and "ultraprocessing," much like "unnatural" and "industrial," as a shorthand for "bad."

What does this mean for nutrition beyond individual consumer choices? Nova, its proponents claim, is meant to be an epidemiological and not a nutritional framework, one that can help experts and regulators identify how particular foods contribute to bad health outcomes. Proponents of Nova don't want sugary breakfast cereals shaping how children eat or sodas to be our go-to drinks. It's a noble goal. What can be done about it?

First, consider that many people eat premade meals, either bought at supermarkets or through food services, which are ultraprocessed. Making those meals healthier, while preserving the convenience factor that leads people to them, would have a much larger public health benefit than trying to convince people to not eat those meals in the first place. A 2023 report by the British food NGO Madre Brava made a case for institutions and consumers calling on food producers to bring their ready-to-eat meal offerings more in line with World Health Organization and EAT-Lancet standards, including using less salt, sugar, and beef, and more vegetables.[37] They'd be more healthy without diminishing the benefits they confer to the consumers who like them and the institutions that rely on them.

Similarly, based on European tax policies, we could replace food sales taxes with "value-added taxes" (VAT) that require producers to pay a tax at every additional stage of processing.[38] Producers could apply for VAT exemptions if, following Fang Fang Zheng's suggestion, the processing provided a nutritious product or, perhaps, safely reduced the perishability of a product. This structure would make the unhealthiest foods more expensive and incentivize producers to reduce processing-for-profit's sake.

Meanwhile, some of the *ingredients* of ultraprocessed foods could quite simply be more strictly regulated. Take soda, which is a perfectly fine guilty pleasure but certainly shouldn't be an ever-present go-to when you're thirsty. There are tried and tested—albeit often politically unpopular—ways of reducing soda consumption. Soda taxes can reduce sugar intake, especially if they're instituted among multiple adjoining municipalities. Removing soda vending machines from schools also reduces soda consumption among young people. Employers, both private and public, should reassess how the food landscapes they maintain within their facilities, from vending machines and cafeteria menus to snack bowls and free soda in the break room, transform what should be exceptional foods into routine foods for their workforces. And although initiatives like these can lead to rancorous debates about consumer freedom, many surveys show that majorities of citizens come to support restrictions on unhealthy foods once they are implemented and their benefits are realized, as was the case with trans-fat bans. But that has little to do with processing in the abstract and everything to do with how we regulate particular, sometimes processed, unhealthy products.

That also applies to ensuring that ostensibly healthy processed foods are, indeed, healthy. Most public-health authorities, including the World Health Organization and the CDC, urge mothers to breastfeed their newborns for their first six months. Many breastfeed for longer than that. But for any number of reasons, ranging from mother's and baby's health to the availability of maternity leave, many mothers supplement or replace breast milk with baby formula. The debate over breastfeeding and formula feeding is a complex one that includes social, economic, cultural, political, and gendered factors, but one main point cuts through it all: For many babies, well-designed formula can be an important part of a healthy diet. Formula, like Huel, is created from extracted vitamins and minerals from other foods to create a delivery mechanism for nutrients optimized to replace or supplement breast milk. Given that it is fed to the most vulnerable among us, its safety should be nonnegotiable. Alas, often it is not.

In 2022 the United States suffered a grievous baby-formula shortage. Lingering supply-chain disruptions from COVID ran headlong into a massive recall by Abbot Nutrition, one of the country's biggest baby-formula producers, after its products were linked to a dead infant and many others sickened by products contaminated with salmonella and *Cronobacter sakazakii*, a rare germ that primarily targets children. The formula was produced in a plant in Sturgis, Michigan, that was beset by leaking roofs and repeatedly cited by the FDA for unsanitary conditions. Not that the FDA was blameless—at the time of the outbreak, it hadn't inspected the plant in two years. In fact, the US government has a history of working on behalf of baby-formula manufacturers to weaken the more stringent standards for baby-formula ingredient safety that American producers face in EU markets. There is no hell hot enough for people who put profit over the health of babies, and this is a far cry from the policies that sought to improve public health through processing. But, yet again, the issue here is the need for stricter regulation and better enforcement, and the political willingness to crack down on bad corporate actors, not the processing itself.[39]

A FIFTEEN-MINUTE WALK NORTH FROM THE FRIEDMAN SCHOOL, NESTLED in Boston Public Market, is a shop called Bagel Guild. Many Bostonians will tell you it's the city's best bagel: peculiarly fluffy and dense at once. Gabriel likes his with cream cheese and lox, while Jan brings his own Tofutti to slather on. We both like our bagel with coffee from the coffee joint just across the market; Gabriel drinks it black, Jan with oat milk. You can plug that bagel and coffee into your MyPlate App most mornings and easily square it with the rest of your day's healthy menu.

But here's the thing about that breakfast you might miss: It is a happy product of modern food technology and industrial processing. The wheat in that bagel is likely a crossbreed, grown as a monocrop, milled in a factory, fortified with minerals and vitamins, and shipped thousands of miles. It's baked in a bakery that follows strict health regulations imposed by the city of Boston and the great state of Massachusetts. The cream cheese? Nova 4. Reconstituted, pasteurized milk product mixed with carob bean gum and potassium sorbate. The lox? Safe packaging and a refrigerated cold chain mean you're not thinking about spoiling and foodborne diseases. Opting for the vegan schmear? Nova 4 too. Soybean oil, soy protein isolate, with a dash of added sugar. The coffee? It's grown in Central America, industrially roasted and processed and shipped in bulk over thousands of miles. The oat milk? Nova 4 again, baby. Oats, rapeseed oil, added vitamins and minerals. Only you don't think about it that way, and you shouldn't. It's delicious, it's healthy, and it's cheap. And you wouldn't think twice about whether it was safe or, indeed, if there would be bagels available today or not. That is because of the unseen beneficence of modern food technology and processing and of the regulatory state.

It's not just that, on balance, we like modern technology and industrial systems. Achieving democratic hedonism for billions of people requires them. Without them, food will be less varied, healthy, affordable, and safe. But much like hedonism needs democracy to be worth defending, so too does food technology. Good technology alone doesn't guarantee good outcomes. How humans use a particular piece of technology (or a particular ingredient) is what matters, and seriously reckoning with use requires that we analyze who gets to decide how technology is used as well as whose interests matter.

Conclusion

FEED THE PEOPLE!

IMAGINE A FAMILY ROAD TRIP THROUGH THE SOUTH A FEW YEARS from now. It's been a long day, including a stop at a pick-your-own-peaches farm north of Athens, Georgia. Now you're sitting in a booth at a Waffle House right across from the grill, with a basket of those peaches on your table. You know you shouldn't, but you can't resist; besides, the waitress said she doesn't mind. She's just working for a tip, you tell yourself, but then you remember there are no tips at the Waffle House. There are even faded stickers on the door that read "No tipping. We pay a living wage." Sticky peach juice running down your fingers, you ask for napkins before you examine the menu. It hasn't changed much from the menu you remember, except for the fact that you can opt for vegan batter for the waffles and mung-bean egg instead of chicken eggs. There's mycelium-based steak and Impossible burgers for the patty melts. Actual red meat is off the menu, as it is in most restaurants across the country after improvements to animal-welfare laws and the implementation of carbon taxes on agriculture.

Your uncle grumbles something about government overreach. Your niece says it's weird to eat cows. They glare at each other over the fast-emptying basket of peaches. You think back to something the peach farmer had mentioned offhand: that next time you're in this neck of the woods you should check out the farm a few miles down the road that had once been an industrial hog operation but was now converted to mushroom cultivation. They let you tour the packinghouse and everything. But better than the mushrooms, she said, is that the stench doesn't waft over on windy days.

Your uncle orders hash browns and eggs. Real eggs, he insists, agreeing to pay the surcharge. Your niece rolls her eyes. She gets the hash browns with the mung-bean egg. You order the pecan waffle because why not. You can't help but watch the guy manning the waffle iron, who moves with the relaxed precision of someone who's done this thousands of times. The back of his black company polo is emblazoned with the number 5: five years as a partner in the Waffle House co-ownership program in the chain of union-staffed, union-managed franchises that now make up most of the restaurant's stores. The waitress brings coffee. It's as average as ever. Her shirt has a union pin on it. Union jobs are hard to find in Georgia, you say, just making conversation. Hey, I just need the benefits in case I have to break up a fight on a Friday night, she says, maybe not joking. Waffle Houses aren't as rowdy as they used to be, but they're still twenty-four-hour diners, after all.

What have we come to, fast-food socialism? jibes your uncle. Actually, you retort, I think you mean fast-food social democracy, if we're splitting hairs. He laughs. Then the food comes, and you all tuck in. The peaches you brought are fresh and juicy, the perfect complement to the greasy deliciousness that this kind of comfort food provides. Then you all sit there, fingers sticky with peach juice and waffle syrup, satisfied, absorbing a moment of respite before hitting the road again. This future is so similar but so different. The same, but better.

* * *

OUR PROMISE OF A BETTER FOOD SYSTEM IS SIMPLE: KEEP WHAT works and get rid of what doesn't.

Here's what works: Food in America is abundant, safe, and cheap, and it has grown markedly more so over the last century. Much of that food is delicious, and the American consumer has more access to a variety of fresh and nutritious foods than at any other time in our history. Our food production is amazingly efficient, with economies of scale leveraging modern technology. It requires relatively little labor, so most people spend their lives doing things besides just growing food to feed themselves. Without the productivity unlocked by the modern food system, Americans would be poorer, sicker, and unhappier. But just as the modern food system makes much that is incredible about our society possible, so too does it reflect the worst parts of our society, from the environment to labor to corporate power to public health. But these problems are far from insurmountable. Some are easier political lifts than others, but none involve a hard reset of the food system.

It's a tricky position to be in favor of large progressive reforms to an existing system while praising many things about that system, to stand on the side of incrementalism rather than abolishing the status quo. Defenders of the system will call our proposed reforms unfeasible or a bridge too far, even as critics will call reformism insufficient and, perhaps, conservative. But the approach of this book, rooted in a food-systems analysis and a pluralistic account of what people want from food, is that we should acknowledge both the changes that need to be made and the limits to those changes. A food-systems approach shows that the food system is complexly composed of myriad interconnected parts and that even small, well-designed interventions can yield large benefits at scale.

Here are some top-line recommendations we hope you'll chew over.

Food in America is highly *available*, more so than in perhaps any other society throughout human history. A major component of that is the basic soundness of our plant agriculture, which, when it contributes directly to human calories and nutrition, is highly efficient and productive, and is, on balance, the least harmful component of our system of food production to the planet and to one another. We should direct public resources toward technologies, such as GMO and gene-edited seed technology, that enhance the efficiency and productivity of the foundation of that system, which can be easily slotted into the existing system and widely adopted at scale. Investments in those technologies should be backed by policies that close "agricultural exception" loopholes in environmental regulations to disincentive the wasteful overproduction of crops for animal feed and ethanol, and that work to retire marginal agricultural land from production and rewild it.

Although the fundamental technology of plant agriculture is highly efficient, the fundamental technology of animal agriculture is inherently inefficient: Protein and calories must be channeled through the metabolic processes of cows or pigs or chickens before they become meat, a process that is both inefficient and harmful (most of all to the animals). People should eat less meat, and, ideally, it would be less available. Policies, including closing agricultural exceptions and imposing taxes that force producers or consumers to internalize costs of production currently borne by society writ large, would help to accomplish that. But because both the scale of current meat consumption and the magnitude of the harms associated with it are so huge, simply working around the edges isn't enough. Long-shot policies that look off-farm, such as public investments in cellular agriculture and alternative proteins that would make it easier for eaters to make dietary changes, have to be on the menu as well. These are the sorts of creative solutions that we will need to make the food system more *sustainable* while preserving the pleasures that it delivers to ordinary people.

Although the food system makes food highly available for most consumers, availability isn't *access*, and millions of Americans struggle with access because they cannot afford the food that is available. Solutions aimed at expanding availability—such as subsidizing the construction of food amenities like groceries, farmers' markets, and community gardens—do not effectively or efficiently expand access, but there are existing programs that do, including SNAP and universal school lunches. Both of these programs are well tailored to address access, but to work at scale they must engage basic components of the conventional, industrial food system like economies-of-scale production and processing of staples. We think that's just fine. Expanding food access is about giving people cash and food, and so programs like SNAP and universal school lunches should be vigorously defended and expanded.

The people who work in the food system, including farmworkers, grocery-shelf stockers, servers, and cooks, are paid miserably low wages and face unsafe working conditions. That should be fixed. Food workers deserve a *fair* meal, and we say that should start with a healthy hike to the federal minimum wage. Beyond that, food workers must be able to organize for better wages and working conditions. That requires strengthening the right of workers to unionize and exploring sectoral bargaining, particularly in corners of the food system where unions alone may not be effective. All of this requires—yes, once again—closing agricultural exceptions and building the capacity of the state institutions that regulate employers and workplaces.

If workers and eaters can afford the bounty our food system makes available, the final question is how we can help them piece together a healthy diet in a world that can make healthy food choices difficult and confusing: How can we make sure that they *utilize* the bounty that is available and accessible? This is one place where individual choice and education, for better or worse, is a key to good outcomes, and there is no substitute for encouraging people to get their dietary advice from experts rather than health and diet influencers. Some simple policies, such as better food labeling and placing supplements under the authority of the FDA, as well as more controversial ones, such as VATs and soda bans, would make it easier for consumers to avoid harmful and calorically dense foods. Meanwhile, ensuring that the people we trust to regulate the food system are up to the task means investing in regulatory capacity, not abandoning it.

THROUGHOUT THIS BOOK, WE HAVE FOREGROUNDED POLICIES THAT will have tangible impacts and have avoided calls for sweeping overhauls of the entire food system that are unlikely to ever be enacted. For instance, changing SNAP eligibility requirements to track cost of living would open the program to a swath of needy Americans, improving their lives and their children's, all for the cost of a rounding error in the federal budget. Those new SNAP recipients would, in all likelihood, still spend their expanded food budgets on conventionally produced food sold at supermarkets and convenience stores. It might change nothing else about the food system, yet it would deliver massive health and social benefits, slashing food insecurity and freeing millions from cycles of hunger and poverty.

Similarly, reducing the amount of meat people eat at a large scale is a much bigger ask, challenging corporate interests, cherished pleasures, and deeply entrenched eating habits, but its effects on the environment would be quite literally world-changing, even if it involved people gobbling a soy-based burger at a fast-food joint or grilling one sold at Walmart. In fact, the environmental benefits of substituting chicken for beef and soybeans for chicken are so substantial that facilitating those substitutions, even through "fast" and "industrial" venues, is among the most consequential ways we can ease the environmental impact of the food system. Internalizing the environmental and ethical costs of animal agriculture through strong regulation and sensible taxation won't do that overnight. Nor will broad-based public investment in alternative protein and food technologies. But stack up enough of these, and the status quo looks less and less like the status quo until, suddenly, you're happily eating mycelium steak at the Waffle House.

If this book asks the reader to change any one thing about how they think about food, it's this: The most important thing you can do is banish from your thoughts the idea that we should—and can—scrap the "broken" food system and start over. And perhaps the best way to do that is to engage in a thought experiment. Let's just imagine that with the wave of a magic wand you can follow those authors we wrote about in Chapter 1 and do away with the Waffle House, fast-food joints, Walmart supercenters, factory farms, large monocropped fields, the USDA, processed junk foods, most modern farming machinery, and so on. Wipe it all clean and start from scratch.

But you're still going to hanker for a waffle. And if you do, you still need wheat. And you need to mill it. And you need the other ingredients to mix it into a batter. Then, if you want to enjoy waffles at something like a family-owned restaurant (or, better yet, a worker-owned waffle cooperative) that buys organic flour only from small farms, cooks it up with only local free-range eggs, tops it with only local butter and real maple syrup, and serves it to local customers, you're going to need to think about how to structure a food value chain that gets waffles onto plates. So surely this already looks vastly different from the current food system, right? Not really. In fact, it might vary in some of the particulars, especially at the start, but you'd need most of the very same structures that, with a wave of your magic wand, you just made disappear.

You'll need to build something like the USDA to ensure that wheat is indeed grown organically according to binding standards, lest every buyer has to inspect every farm and every sack of grain. What about protecting farmers from bad crops or droughts and buyers from losing their supply in such a case? Congrats! You just reinvented the Federal Crop Insurance Corporation and the Commodity Futures Trading Commission. And unless you're relying on the benevolence of farmers and restaurant bosses, work hours, wages, and labor safety seem to call for codified legal standards and protections. Ah, we see you've just reinvented the Department of Labor and OSHA. What about the safety of that milk and those eggs in the batter? That's called the FDA.

Too many people who value a functioning commonsense regulatory state when it comes to car and drug manufacturing turn into Ayn Rand if you're talking brussels sprouts, avocados, and tenderloins. But the fact, as we hope we've made clear throughout the book, is that the regulatory state is inherently and by necessity a part of a well-functioning food system. Standards, rules, laws, oversight—these fundamentally modern and industrial principles are the bedrock of any food system that works reliably at scale.

Then, of course, comes the economic feasibility of your new food system. Where are the fridge, freezer, cooktop, and waffle iron for the restaurant going to come from? You'll need a new manufacturing supply chain to make the specialized appliances that likely won't be produced by a mom-and-pop manufacturer. And all that's before we get into the logistics of producing and sourcing enough ingredients for such bespoke waffle restaurants, much less something as simple as bread, in major cities like New York, Chicago, or Atlanta, and the small army of warehouses and drivers and other middlemen like grocery stores that this would require. And then questions remain about integrating these small farms with SNAP and food banks and school lunches, unless we just do away with those as well when we scrap the food system. We could go on, but we think the point we're making should be becoming obvious.

As you try to work out the components of a replacement food system, you end up re-creating many of the foundational parts of the current one. And that's not a coincidence. We can't have a modern world without a modern food system. And being crystal clear about this is the first step in working toward building a better one.

SPEAKING OF REALISM, HERE'S ONE POLITICAL REALITY WE MUST observe: Food has to stay affordable for consumers, and agriculture has to continue to be highly productive. Less productivity means more land use or more reliance on labor in dangerous and poorly compensated jobs. More expensive food will increase food insecurity and could cause a further deterioration in nutrition. Throughout this book we have argued for practices and policies that preserve and improve productivity and explained how that in turn can shore up support for progressive practices and policies. We also recognize that although workers and communities sometimes reap the benefits of improved labor productivity in the food system, it's often been siphoned off by the already rich and powerful, leaving problems to fester and worsen. Part of building a better food system means recapturing the gains from highly productive modern agriculture and redistributing it to those who need it most or redirecting it away from harmful ends to beneficial ones, be it for people or for the planet.

Potential changes to the food system should be made with an eye toward increased productivity on the one hand and affordable and abundant food on the other. But these are not worthy ends in and of themselves. Low food prices are, in general, a good thing. They are not only politically important, but they are also a great outcome of modern food production because they fulfill a basic human need—food—at the cost of minimal labor. But pushing for low prices isn't worth putting kids to work cleaning meatpacking plants or gutting the minimum wage. Not every bit of productivity savings is worth it. The whole point of keeping food affordable is to improve general social welfare, not to immiserate some portion of the workforce just so that a food corporation's stock price perks up a few extra cents. Put differently, we are so emphatically supportive of laborsaving technological improvements guided by public investment and regulation precisely because they can ease the cost pressures that higher labor costs might incur.

Nor do we think *all* food should be as cheap as possible. In fact, some foods should be more expensive to disincentivize their production and consumption, such as sugary drinks because of their health effects and red meat because of its contribution to climate change. But the principle that a nutritious diet is accessible to all people should guide visions for a better food system and, indeed, may be the number-one demand the public makes of those who would change it. We remain skeptical about the efficacy and political feasibility of outright bans on foods; thus, we favor regulations that are consistent with the dignity of individual choice, even as we are candid about the need for all food items to bear the true social costs of their production and consumption.

Meanwhile, embracing productivity for its own sake is part of the reason we have many of the problems that we do with the food system. US corn agriculture is enormously productive, so much so that we've found useless things to do with all that corn. Being productive is good if you're producing good things (like delicious affordable food), less if you're producing toxic manure lagoons and biofuels. Ignoring this basic difference can lead to outright outlandish suggestions, such as a *New York Times* op-ed published in December 2024 that argued for more factory farms because they are the most efficient way of producing meat and because the author seems overly certain that people will never reduce their meat consumption—an assumption often made but rarely substantiated.[1] This sort of narrow-minded thinking about productivity got us the horror of factory farms in the first place and is also why so much land is wasted growing corn to burn as ethanol. When we argue for productivity, it is in the sense of using the tools and technologies of modern agriculture and applying them to the general task of creating an abundant and affordable supply of food for direct human consumption.

Meanwhile, we commend treating with respect and trying to understand other people's pleasures—yes, even their meaty pleasures—but with the proviso that we also have to be candid with people about the scientifically documented harmful consequences those pleasures bring. Democratic hedonism's commitment to pleasure pluralism comes with a healthy dose of being willing to tell it like it is. As we discussed in Chapter 3, CAFOs are a relatively more efficient way of producing animals, yes, but a very inefficient way of producing calories and protein at scale. It is, in the aggregate, *un*productive. You can treat people who eat meat with respect while being clear about this fact; indeed, being honest in your disagreement is what real respect looks like.

Continuing to improve productivity in food and agriculture requires an openness to technological innovation that will not square with the idea that farming can, or should, be returned to a simpler, purer Edenic state. The surest way to ameliorate many of the most grievous harms generated by the food system—even those allegedly caused by the food system's supposed industrialization—is more and better technology. Want less pesticides on your food, for instance? The most reliable ways to reduce pesticide use at scale are better spraying technology and GM varieties that are engineered to be more resistant to pests.

Investing in technologies that can improve the food system is an excellent use of public resources. Historically, agricultural science has been bathed in public resources through the Department of Agriculture, the Cooperative Extension Service, and the land-grant colleges. That's a good thing, and it should continue, but with a view toward building a more sustainable food system rather than padding the pockets of incumbent players. That means shifting funding from programs that, for instance, want to squeeze a little more meat out of factory-farmed animals to ones like Tufts University's USDA-funded

Center for Cellular Agriculture that is trying to develop promising technologies. There is currently no national food science and technology laboratory, for example, and much of the basic science that goes into our food is guarded by multinational corporations and their armies of white-shoe attorneys as intellectual property. Creating and generously funding such a laboratory is essential, but so is making sure that the knowledge and technology it produces are publicly owned and widely disseminated.

New technologies, even those that will definitely improve the food system, are not, however, always an unmitigated boon for society. They can be used to exploit people. They may help to line the pockets of unsavory monopolists. Even the most promising technologies must be guided and regulated in the public interest. Processing technologies—such as those that have enriched flour and made baby formula possible—can make food better (precisely why we believe using "processed" as shorthand for bad is misguided). But when food manufacturers use processing to render foods addictively "hyperpalatable," they are guided by private profit, not public welfare. Processing as a whole is not the problem. But some processed foods are.

Scrutiny and regulation of new technology require serious regulatory capacity and well-designed institutions capable of monitoring large-scale problems and implementing large-scale solutions. One problem with the nostalgic, smaller-is-better line of foodie thought is that it is usually allergic to large-scale institutional thinking and design. But even as those people would rather stick their heads in the fertilizer pile of an imagined agrarian past than find large-scale solutions to the problems of the present, they are frequently right about one thing: The sad truth is that the institutions entrusted with regulating food and agriculture in the United States are currently not doing their job all that well. Restoring a measure of faith in those institutions is critical, but it requires engaging with and reviving them.

We would start with the failings of the Department of Agriculture, one of the recurring sources of both hope and disappointment in this book. The USDA is a gargantuan public agency entrusted with regulating agriculture in the name of the public's welfare and safety. But it's also agriculture's designated booster, which means that it is expected to make sure that farmers and agribusinesses are making money. When the USDA was founded, in 1862, those two missions could be squared—farmers were still a majority of the public, so the nation's well-being depended on theirs. Today, farmers are a minuscule fraction of the population. That's generated not only misalignment but incompatibility in the USDA's fundamental responsibilities: How can you look after the profitability of large farms and multinational corporations when those entities make their money exploiting the very public you're meant to be protecting?

A more sensible design would clarify the USDA's mission and restructure the agency to effectively pursue it. Regulators and boosters should be separated. And the agency's primary mission, rather than the profitability of farmers and agribusiness, should be to cultivate the public's reliable access to sustainable, affordable, delicious, and nutritious food. In other words, the USDA should be accountable to the public and only the public. We'd signal that new mandate by renaming it "The Department of Food." And it should focus on the biggest parts of its budget that serve those ends: SNAP and school lunches. To the extent that farmers are supported, they should be supported for growing items like fruit and vegetables rather than monocrop corn for animal feed and biofuels. This, in turn, might mean supporting some smaller farms here and there, and could even be paired with efforts to support new farmers and those who come from historically excluded populations.

A better USDA would trailblaze, rather than hinder, more sustainable uses for land currently dedicated to growing crops for animal feed and ethanol. The biggest environmental benefits of changing land-use policy come when agricultural land is not only retired but also returned to wilderness (or as close as we can come to it). Large-scale rewilding initiatives would take serious government investment and oversight through existing land-conservation programs or new land banks, but their environmental and social benefits could be astronomical.

It's not all the fault of the USDA's poor design. To the extent that regulators do police agriculture, it's with far weaker tools than in other industries. Agriculture enjoys broad carve-outs and exemptions from many aspects of the modern regulatory state. In this book we've mentioned "agricultural exceptions" to laws regulating air and water pollution, animal cruelty, collective bargaining, workplace health and safety, and even (it still disgusts us to write this) child labor. Legislators should plug the holes. Farms should operate according to the same high environmental, safety, and labor standards by which all other businesses must abide.

Agricultural exceptionalism is just a start. Bad laws run up and down the food system. On the farm, the subsidization of commodity corn is partially a product of federal laws designed to encourage bio-fuel use. Off the farm, the miserly federal minimum wage of $7.25 per hour that immiserates food service workers, as well as the exemption of tipped workers from that minimum, is a product of federal law. Federal law is similarly to blame for the understaffing and underfunding that inhibits the ability of the National Labor Relations Board to keep pace with the rising tide of union organization at restaurants, coffee shops, and grocery stores.

These laws should be changed, new laws should be added, and existing laws should be enforced. To put it as plainly as possible: A good food system requires a robust regulatory state, one that attends to everything from ensuring fair competition to food safety to compliance with environmental and labor standards. The public and lawmakers also need to gird themselves against the industry's one messaging trump card when it comes to any regulation: Prices will go up. Yet, as we've shown, this simply isn't the case. And if *some* products can't be produced profitably while adhering to fair laws and regulations, maybe those products shouldn't be part of our food system.

In that vein, taxes on carbon emissions and a system of carbon credits would positively transform the food system from top to bottom, encouraging energy efficiency, sustainability, and productivity. A well-designed value-added tax on food products would apply a counterweight to needless and extractive processing and would be a great complement to a revamped agricultural subsidy regime that actually prioritized the affordability of fresh fruits and vegetables.

Federal antitrust litigation should be used to break up the handful of swollen agribusinesses that monopolize the food system, stimulating more competition. And the same antitrust measures should bring serious scrutiny to consolidation of grocery stores and other food retailers. Kudos to the Federal Trade Commission, for instance, for blocking Kroger's planned $24.6 billion acquisition of Albertson's in late 2024. Still, trust-busting can take us only so far. Even competitive markets have big, industrial-scale players, and the same will be true in food and agriculture. To keep the big guys in line, we also need new, stronger laws; better staffing; and more funding for our Department of Food, the Department of Labor, OSHA, the EPA, and other key regulatory agencies to make sure that jobs in the food system are dignified, safe, and fairly compensated. Contrary to the DOGE mindset, investing in government institutions will help eaters, workers, and innovators alike.

BIG INSTITUTIONAL, REGULATORY, AND SYSTEMIC PROBLEMS ARE JUST that: big. They are large-scale and long-term, measured in terms that sometimes outstrip our ability to think about them. You might be shocked by the gruesome, painful death of a single animal and decline to eat the bacon made from its body, but the prospect of ten billion animals? People might be shocked to hear that number but still eat the bacon. After all, what does it really matter? Similarly, when you see how hard your Waffle House server is busting their ass for that $3-per-hour wage, you may be moved to put down a much juicer tip. But 22 million people being paid terrible wages to feed us? You can't tip 22 million people out of poverty, and the battle over minimum wage reform may seem either pointless or just so far away from your other everyday concerns.

It's not just that the scale is hard to think about. Thinking about, much less doing anything about, big problems *feels* scary, daunting, and off-putting because it reminds us of how small we are by comparison. And that's a struggle not just when it comes to food, but a basic reality of the complexity and interconnectedness of the world in which we live—a challenge of modern politics in the Information Age that none of us has yet fully digested. It's easy to leap from those feelings to two faulty ideas about what you, as an individual, can do about all this.

The first bad idea we want to caution you against is the notion that there's a rigid boundary between individual actions and collective actions, and that the existence of the latter relieves you from having to worry about the former. Collective actions, whether political, economic, or cultural, are composed of individual actions, and they blend together seamlessly in the interlaced, heavily mediated social reality we inhabit. Observing that a problem is systemic, or structural, or just really, really, really huge—that's not a reason to ignore how we, as individuals, retain agency and responsibility to positively transform our world where we can. You're not wrong to pass on the bacon or tip your server. The fact that these individual acts don't solve the totality of huge problems is not a good reason to shirk doing your part when you can. As with anything when it comes to the complexity of the food system, our inability to solve everything all at once is a poor excuse for declining to fix what we can, where we can, when we can.

The second bad idea is the flip side of this: that the correct direction of concern for your agency is exclusively inward-looking. Although we've got, obviously, nothing against local activism and individual actions, too much thinking local can sometimes produce a downward and recursive spiral. The most local things of all, after all, are your body and your neurotic anxieties about it. Bodies and neuroses—those are the tireless focus of the diet huckster, regeneration evangelist, supplements peddler, anti-vaxxer-curious raw-milk enthusiast, libertarian homesteader, body hacker, carnivore guru, and all the other malignant con men who have, over the years, spilled out of the food-influencer clown car with big plans (and scant peer-reviewed studies) for your personal consumption habits.

Diet and nutrition, for all the obscurantism, smoke, and mirrors of diet discourse, don't benefit much from obsessive modulation. In fact, we think peer-reviewed science provides simple guidelines for dietary habits that will keep most people reading this book relatively healthy, well-fed, and pleasantly sated: Eat food, including nutritious processed foods. Not too much. Mostly plants. Beyond this mantra, we'd add that monitoring and reducing salt, sugar, and fat intake is likely wise, and something you should absolutely do if your doctor tells you to. But beyond those basics, we'd say there is a very low upside to baroque diet schemes that require obsessive attention and costs to follow.

Once we've moved past the biomedical basics, we can pay more attention to why food is a good portal to political action aimed at helping *other* people and the world beyond ourselves. The pleasures of eating link us to one another. They help us understand that although we may sometimes feel solitary and alone, bereft and abandoned to the big problems of the food system, every time we eat we are connected to other people, and we can viscerally experience our interconnectedness in every bite and morsel. Ask yourself: Who feeds you? What do you owe them? How can you help ensure they're treated fairly? And also:

Whom do you feed? What do you owe them? How can you help them eat better? Finding answers to these questions is the political promise of food and precisely what separates democratic hedonism from mere hedonism: Food, and the pleasure we take from it, draws us out, connects us, makes us feel less lonely, nourishes us, and steels us for all the difficult work ahead.

Yes, there are many things only a robust regulatory state can do. But there are things you *can* do. And you should.

Contact your state or city elected officials, and push for universal school lunches if your community doesn't have them (and demand they build more affordable housing while you're at it).

Request that your university, workplace, or professional association make plant-based meals the default menu options at all official functions.

Ask the mayor why the meal offerings in the courthouse are so dreadful. Shouldn't citizens who show up for jury duty, or to contest a speeding ticket, or who happen to work in the filing room be able to find a nutritious and tasty meal?

Ask your congressperson why hundreds of thousands of children still work on American farms.

Volunteer at your local food bank or migrant shelter; when you donate to it, write a check instead of dropping off your old canned goods.

Tip your waitress and your delivery guy, but also support Fight for $15.

Find out who the Coalition of Immokalee Workers is boycotting right now, and write a letter to the company letting them know you won't cross that picket line.

Don't cross a picket line, not even for a delicious waffle.

Vote for candidates (and ballot initiatives) who will help to make sure that the price of food reflects the true costs of production as well as those that support robust public funding for basic and open access agricultural and food-science research.

Skip the burger when you can, unless it's plant based, and tell your favorite doner kebab guy, the one who always calls you "boss," that you'll happily order a seitan doner if he puts it on the menu.

None of this alone is sufficient, but all of it can make a difference, and you'll feel better for doing it.

Wendell Berry famously called eating an agricultural act. He was right about that, but we'd add it's also a political act. Alas, Berry birthed a food politics that was too cramped and nostalgic to reckon with the full significance of eating and feeding in the modern world, to whom and what it might connect us, and what sorts of commitments we should cultivate to improve it.

We need a politics of eating that engages the industrial, large-scale, and increasingly global horizons of the food system and that sidesteps the stale dichotomy between small and big, individual and collective. It's a politics that will rebuild institutions and write laws that can ease the anxiety and burden that individuals understandably experience when they try to imagine vast problems at large scales, much less implementing solutions to these problems. But it's also a joyful and hungry orientation to food for each of us, and one that will lead us out into the world well beyond our kitchens, doorsteps, garden, and barnyards. When we go into that world, we will still be curious about new dishes yet to be tasted, about new people to meet and break bread with. We will be excited anew to be fed and to feed. A world of delicious promise awaits.

Acknowledgments

This project was perhaps envisioned earlier and more clearly by our agent Eric Henney at Brockman, Inc., than by ourselves, and we are grateful to him for his vision, motivation, patience, and feedback as we researched and wrote this book. The initial ideas for many of the themes discussed here were explored across a dozen articles we cowrote for *The New Republic*, *Vox*, *The Guardian*, and *Dissent*. We are grateful to our editors at these publications, and especially to Heather Souvaine Horn at *The New Republic*, for giving us the intellectual space to slay some sacred cows in contemporary food writing. At Basic Books, Emily Taber was a reliable voice of order, reason, and realism. Julieta Cardenas and Indigo Oliver were invaluable research assistants at different stages of this project.

Duke University, the National Humanities Center, and the Max Planck Institute for the History of Science generously provided Gabriel with the resources and time that made this book possible. Gabriel is especially grateful to Ben Miller, Tamar Novick, Lisa Onaga, Dagmar Schäfer, and Ben Trott, for helping to make Berlin his second intellectual home. His colleagues at Duke University, especially those in the Program in Gender, Sexuality, and Feminist Studies and History, have been stalwart intellectual comrades and great friends, especially Nima Bassiri, Samuel Fury Childs Daly, Adam Mestyan, Jennifer Nash, Priscilla Wald, and Kathi Weeks. Gabriel owes more to Harris Solomon than words can express. Many years ago, Gabriel was a postdoctoral fellow in Yale University's Program in Agrarian Studies. There, the wisdom and humanity of the late James C. Scott shaped his thinking about agriculture. Several years later, Gabriel and James organized a conference on the ten-thousand-year history of hogs in human societies, at which Gabriel met a young Polish political scientist from the New School for Social Research who presented a paper on welfare standards on contemporary hog farms. The rest, as they say, is history.

Jan's research on the food system that set the groundwork for this book has been helped by extremely supportive research environments and colleagues over the years. He is especially grateful to Jessica Pisano and the Department of Politics at the New School for Social Research, Adam Sheingate and the Department of Political Science at Johns Hopkins University, Chris Green and the Animal Law and Policy Clinic at Harvard Law School, and his colleagues at the Pratt Institute. And he is most grateful, as always, to Ewa, Piotr, Tania, and Vera.

Notes

Introduction: The Food System Paradox

1. *Anthony Bourdain: Parts Unknown*, season 6, episode 8, "Charleston, S.C.," directed by Tom Vitale, featuring Anthony Bourdain and Bill Murray, aired November 15, 2015, on CNN.

2. Fiza Pirani, "Waffle House, by the Numbers," *Atlanta Journal-Constitution*, August 13, 2019, www.ajc.com/entertainment/dining/waffle-house-the-numbers/nfNge1HkMGFGY1hWgEQwBM.

3. Food and Agriculture Organization of the United Nations and United Nations Environment Programme, *The State of the World's Forests 2020: Forests, Biodiversity, and People* (FAO and UNEP, 2020).

4. Matthew P. Rabbitt et al., *Household Food Security in the United States in 2022*, Economic Research Report no. 325 (US Department of Agriculture, Economic Research Service, October 2023), https://doi.org/10.32747/2023.8134351.ers; Brady Stierman et al., *National Health and Nutrition Examination Survey 2017–March 2020 Prepandemic Data Files: Development of Files and Prevalence Estimates for Selected Health Outcomes*, National Health Statistics Reports no. 158 (National Center for Health Statistics, 2021); Centers for Disease Control and Prevention, *National Diabetes Statistics Report, 2022*, www.cdc.gov/diabetes/data/statistics-report/index.html. On agriculture as a driver of zoonotic illness, see M. N. Hayek, "The Infectious Disease Trap of Animal Agriculture," *Science Advances* 8, no. 44 (November 4, 2022).

5. Michael Pollan, *The Omnivore's Dilemma: A Natural History of Four Meals* (Penguin, 2006); Alice Waters, *We Are What We Eat: A Slow Food Manifesto* (Penguin, 2021); Wendell Berry, *The Unsettling of America: Culture and Agriculture* (Sierra Club Books, 1977).

6. Nourish, "Food System Tools," accessed June 30, 2025, www.nourishlife.org/teach/food-system-tools.

7. On wheat-land use statistics in Kansas and North Dakota, see "2024 State Agriculture Overview: Kansas," US Department of Agriculture, National Agricultural Statistics Service, www.nass.usda.gov/Quick_Stats/Ag_Overview/stateOverview.php?state=KANSAS; and "2024 State Agriculture Overview: North Dakota," US Department of Agriculture, National

Agricultural Statistics Service, www.nass.usda.gov/Quick_Stats/Ag_Overview/stateOverview .php?state=NORTH+DAKOTA. For histories of wheat cultivation attentive to biology, see Catherine Zabinski, *Amber Waves: The Extraordinary Biography of Wheat, from Wild Grass to World Megacrop* (University of Chicago Press, 2020); and Scott Reynolds Nelson, *Oceans of Grain: How American Wheat Remade the World* (Basic Books, 2022). On wheat farming on the Great Plains in the twentieth century, see Paul K. Conkin, *A Revolution Down on the Farm: The Transformation of American Agriculture Since 1929* (University Press of Kentucky, 2008); and R. Douglas Hurt, *Problems of Plenty: The American Farmer in the Twentieth Century* (Ivan R. Dee, 2002). On the Ogallala Aquifer, see Lucas Bessire, *Running Out: In Search of Water on the High Plains* (Princeton University Press, 2021).

8. On the history and business models of Monsanto and ADM, two of the most significant agribusinesses, see Bartow J. Elmore, *Seed Money: Monsanto's Past and Our Food Future* (W. W. Norton, 2021); and James B. Lieber, *Rats in the Grain: The Dirty Tricks and Trials of Archer Daniels Midland, the Supermarket to the World* (Basic Books, 2000).

9. On Waffle House's political contributions, see "Waffle House Inc Profile: Summary, 2022 Cycle," *OpenSecrets*, accessed March 30, 2025, www.opensecrets.org/orgs/waffle-house -inc/recipients?id=D000021946&cycle=2022.

10. Rudd Center for Food Policy & Health, *Fast Food FACTS 2021: Billions in Spending, Continued High Exposure by Youth* (University of Connecticut, 2021).

Chapter 1: The Case for Democratic Hedonism

1. Alice Waters, introduction to Wendell Berry, "The Pleasures of Eating," *Emergence Magazine*, September 30, 2019, https://emergencemagazine.org/essay/the-pleasures-of-eating; Michael Pollan, "Wendell Berry's Wisdom," *The Nation*, September 21, 2009, 25–28.

2. David Derbyshire, "Wine-Tasting: It's Junk Science," *Guardian*, June 23, 2013.

3. Calvin Trillin, "The Red and the White," *New Yorker*, August 11, 2002.

4. Ali Francis, "The Disgraced Fine Dining Restaurant Willows Inn Has Closed," *Bon Appétit*, December 5, 2022, www.bonappetit.com/story/willows-inn-closed-allegations-abuse.

5. Joseph J. Fischel, "Toward a Democratic Hedonism," *Boston Review*, May 20, 2019, www.bostonreview.net/articles/joseph-j-fischel-toward-democratic-hedonism.

6. All quotations of Berry in this section come from the version of the essay published in Wendell Berry, "The Pleasures of Eating," in *What Are People For?* (North Point, 1990), 145–152.

7. BELCAMPO INFO (@nela_butcher) "Taken from my story. Please repost to help spread awareness of this fraud of a company. These types of practices have gone on for way too long in the meat industry, it's time for change. Thank you for your support 🐑," Instagram, May 26, 2021, www.instagram.com/p/CPWOV2pHswY/?utm_source=ig_embed&ig_rid =65bf18b9-53ac-4724-b61d-d047d404f5ea.

8. Stephanie Brejo, "After Sourcing Scandal, Belcampo Meat Co. Abruptly Closes Stores, Restaurants," *Los Angeles Times*, October 19, 2021.

9. "Home," Blue Hill Farm, accessed March 30, 2025, www.bluehillfarm.com; "Home," Chez Panisse, accessed March 30, 2025, www.chezpanisse.com.

10. All quotations of Berry in this section come from Wendell Berry, *The Unsettling of America: Culture and Agriculture* (Sierra Club Books, 1977).

11. Michael Pollan, "Introduction: Our National Eating Disorder," in *The Omnivore's Dilemma: A Natural History of Four Meals* (Penguin, 2006), 1.

12. Pollan, "Wendell Berry's Wisdom," 26.

13. Donald Worster, *Dust Bowl: The Southern Plains in the 1930s* (Oxford University Press, 1979).

14. Steven Stoll, *Larding the Lean Earth: Soil and Society in Nineteenth-Century America* (Hill and Wang, 2002).

15. Emily Pawley, *The Nature of the Future: Agriculture, Science, and Capitalism in the Antebellum North* (University of Chicago Press, 2020); Ariel Ron, *Grassroots Leviathan: Agricultural Reform and the Rural North in the Slaveholding Republic* (Johns Hopkins University Press, 2020).

16. William Jennings Bryan, "The Cross of Gold Speech," July 9, 1896, reprinted in *Teaching American History*, March 30, 2025, https://teachingamericanhistory.org/document/the-cross-of-gold-speech.

17. Chris Smaje, *A Small Farm Future: Making the Case for a Society Built Around Local Economies, Self-Provisioning, Agricultural Diversity, and a Shared Earth* (Chelsea Green, 2020).

18. Mark Bittman, *Animal, Vegetable, Junk: A History of Food, from Sustainable to Suicidal* (Harvest, 2021).

19. Arlie Hochschild, with Anne Machung, *The Second Shift: Working Families and the Revolution at Home* (Penguin, 2012).

20. Martha Nussbaum, *Creating Capabilities: The Human Development Approach* (Harvard University Press, 2011).

21. Eve Kosofsky Sedgwick, *Epistemology of the Closet* (University of California Press, 1990), 22.

Chapter 2: Farming Without Sentimentality

1. Jane Mt. Pleasant, "The Paradox of Plows and Productivity: An Agronomic Comparison of Cereal Grain Production Under Iroquois Hoe Culture and European Plow Culture in the Seventeenth and Eighteenth Centuries," *Agricultural History* 85, no. 4 (2011): 460–492.

2. On the transition from Native American farming to commercial-oriented settled agriculture in the nineteenth century, see R. Douglas Hurt, *American Agriculture: A Brief History*, rev. ed. (Purdue University Press, 2002); and David B. Danbom, *Sod Busting: How Families Made Farms on the Nineteenth-Century Plains* (Johns Hopkins University Press, 2014).

3. See Steven Biel, *American Gothic: A Life of America's Most Famous Painting* (W. W. Norton, 2005).

4. "Century and Heritage Farm Program," Iowa Department of Agriculture and Land Stewardship, accessed March 30, 2025, https://iowaagriculture.gov/century-and-heritage-farm-program.

5. US Department of Agriculture, National Agricultural Statistics Service, *2022 Census of Agriculture: Iowa, State Level Data*, volume 1, chapter 1, April 2024, www.nass.usda.gov/Publications/AgCensus/2022/Full_Report/Volume_1,_Chapter_1_State_Level/Iowa; NASS+5NASS and Iowa Pork Producers Association, "2020 Iowa Pork Industry Facts," www.iowapork.org/newsroom/facts-about-iowa-pork-production.

6. Michaelyn Mankel, "Beyond Voluntary: It's Time to Get Serious About Clean Water in Iowa," *Des Moines Register*, February 12, 2024, www.desmoinesregister.com/story/opinion/columnists/iowa-view/2024/02/12/hold-factory-farming-accountable-for-polluting-iowa-water/72566782007.

7. US Department of Agriculture, National Agricultural Statistics Service, *Agricultural Resource Management Survey*, December 2024.

8. US Department of Agriculture, National Agricultural Statistics Service, *2022 Census of Agriculture*, February 2024, 7; US Department of Agriculture, National Agricultural Statistics Service, *2022 Census of Agriculture: Iowa, State Level Data*, volume 1, chapter 1, table 1, 3; James M. MacDonald et al., *Farm Size and the Organization of U.S. Crop Farming*, Economic Research Report no. 152 (US Department of Agriculture, Economic Research Service, August 2013), 4; US Department of Agriculture, National Agricultural Statistics Service, *Agricultural Resource Management Survey*.

9. For the number of farmers, see the US Department of Agriculture, Economic Research Service, "Farm Labor," last updated January 8, 2025, www.ers.usda.gov/topics/farm-economy/farm -labor. For farmworker jobs, we're going with the Bureau of Labor Statistics official number, but for complicated reasons it may be a dramatic undercount. See Daniel Costa, "How Many Farmworkers Are Employed in the United States?," Working Economics Blog, Economic Policy Institute, October 3, 2023, www.epi.org/blog/how-many-farmworkers-are-employed-in-the-united-states.

10. Austin Frerick, "'Barons' Examines US Food System, Including Pork Production," *Des Moines Register*, March 31, 2024; Charlie Mitchell and Austin Frerick, "The Hog Barons," *Vox*, April 19, 2021.

11. John H. Davis and Ray A. Goldberg, *A Concept of Agribusiness* (Harvard University, 1957).

12. ADM has seen a recent decline in revenue, to $85 billion in 2024. "Archer Daniels Midland Co. Annual Income Statement," *Wall Street Journal*, Markets & Finance, accessed March 30, 2025, www.wsj.com/market-data/quotes/ADM/financials/annual/income-statement; "ADM: Human, Pet and Animal Nutrition Company," Archer-Daniels-Midland Company, accessed April 4, 2025, www.adm.com/en-us.

13. Steven Ramsey et al., *Global Demand for Fuel Ethanol Through 2030*, BIO-05 (US Department of Agriculture, Economic Research Service, February 2023), iv; Taxpayers for Common Sense, *Understanding U.S. Corn Ethanol and Other Corn-Based Biofuels Subsidies*, May 7, 2021, www.taxpayer.net/energy-natural-resources/understanding-u-s-corn-ethanol -and-other-corn-based-biofuels-subsidies.

14. Courtney Leeper Girgis, "Pork Powerhouses 2023: Sow Numbers, Pig Profits Down While Productivity Up," *Successful Farming*, May 6, 2024.

15. James M. MacDonald et al., "Concentration and Competition in U.S. Agribusiness (Report No. EIB-256)," *US Department of Agriculture, Economic Research Service*, June 27, 2023, https://doi.org/10.32747/2023.8054022.ers.

16. James M. MacDonald, "Concentration in U.S. Meatpacking Industry and How It Affects Competition and Cattle Prices," *Amber Waves*, January 25, 2024, www.ers.usda .gov/amber-waves/2024/january/concentration-in-u-s-meatpacking-industry-and-how -it-affects-competition-and-cattle-prices.

17. Austin Frerick, *Barons: Money, Power, and the Corruption of America's Food Industry* (Island, 2024).

18. "A CAFO Turned Specialty Mushroom Farm," 1100 Farm, accessed April 4, 2025, www.1100farm.com.

19. See Scott Reynolds Nelson, *Oceans of Grain: How American Wheat Remade the World* (Basic Books, 2022).

20. G. Stanhill, "Trends and Deviations in the Yield of the English Wheat Crop During the Last 750 Years," *Agro-Ecosystems* 3 (1976): 1–10, https://doi.org/10.1016/0304 -3746(76)90096-2.

21. Charles C. Mann, "Breakfast for Eight Billion," *New Atlantis*, Winter 2025, www .thenewatlantis.com/publications/how-agriculture-system-works.

22. Kent Thiesse, "USDA Crop Report Projects Record U.S. Yields," *Agweek*, September 19, 2024, www.agweek.com/opinion/usda-crop-report-projects-record-u-s-yields.

23. Claudia Flavell-While, "Fritz Haber and Carl Bosch—Feed the World," *Chemical Engineer*, March 1, 2010, www.thechemicalengineer.com/features/cewctw-fritz-haber-and-carl-bosch-feed-the-world.

24. Sarah Garland and Helen Anne Curry, "Turning Promise into Practice: Crop Biotechnology for Increasing Genetic Diversity and Climate Resilience," *PLoS Biology* 20, no. 7 (2022): e3001716.

25. See also Sarah Garland, "Beyond the Shadow of Roundup Ready," *Ambrook Research*, January 11, 2025, https://ambrook.com/offrange/perspective/beyond-roundup-readys-shadow.

26. Ariel Ron, "The Iron Farm Bill," *Phenomenal World*, May 2, 2024, www.phenomenalworld.org/analysis/the-iron-farm-bill.

27. Robert Paarlberg, *Resetting the Table* (Penguin, 2021).

28. Ed Regis, "The True Story of the Genetically Modified Superfood That Almost Saved Millions," *Foreign Policy*, October 17, 2019; Robin McKie, "'A Catastrophe': Greenpeace Blocks Planting of 'Lifesaving' Golden Rice," *Guardian*, May 25, 2024.

29. Rebecca Heilweil, "The Controversy over Bill Gates Becoming the Largest Private Farmland Owner in the US," *Vox*, June 11, 2021.

30. Mike Cernovich (@Cernovich), "'Eat bugs, live in a pod,' is not a meme. It's their real agenda," X, March 25, 2023, https://x.com/Cernovich/status/1639677891716988928; "Watching Too Much Television," *The Sopranos*, season 4, episode 7, directed by John Patterson, written by Terence Winter and Nick Santora, aired October 27, 2002, on HBO.

31. US Department of Agriculture, National Agricultural Statistics Service, "Farm Producers," *2022 Census of Agriculture Highlights*, March 2024, www.nass.usda.gov/Publications/Highlights/2024/Census22_HL_FarmProducers_FINAL.pdf. A much higher percentage of primary operators are men.

32. Gabriel N. Rosenberg, "Inventing the Family Farm: Towards a History of Rural Heterosexuality," *NOTCHES: (Re)Marks on the History of Sexuality*, February 18, 2016, https://notchesblog.com/2016/02/18/inventing-the-family-farm-towards-a-history-of-rural-heterosexuality.

33. Erin Jordan, "Who Owns Iowa Farmland? In Many Cases, It's Not Farmers," *Investigate Midwest*, June 12, 2023, https://investigatemidwest.org/2023/06/12/who-owns-iowa-farmland-in-many-cases-its-not-farmers. See also Donnelle Eller, "Nearly 60 Percent of Iowa Farmland Owners Don't Farm; One-Third Have No Ag Experience," *Des Moines Register*, June 28, 2018, www.desmoinesregister.com/story/money/agriculture/2018/06/28/iowa-state-isu-farmland-farm-facts-ownership-tenure-survey-owners-debt-land-rent-family-income/742159002.

34. US Department of Agriculture, National Agricultural Statistics Service, "Land Values 2024 Summary," August 2, 2024, www.nass.usda.gov/Publications/Calendar/calendar-landing.php?day=02&month=08&report_id=17019&source=n&year=18.

35. David Noyce, "Church's Iowa Farm Team," *Salt Lake Tribune*, November 2, 2023, www.sltrib.com/religion/2023/11/02/latest-mormon-land-church-buys-up.

36. US Department of Agriculture, Economic Research Service, "Income and Wealth in Context," *Farm Household Well-Being*, updated January 27, 2025, www.ers.usda.gov/topics/farm-economy/farm-household-well-being/income-and-wealth-in-context. See also Nathan Rosenberg and Bryce Stucki, "Don't Trust the Antitrust Narrative: Farmers Benefit from

Industrial Ag. Workers Do Not," *Jacobin*, June 9, 2021, https://jacobin.com/2021/06/anti
-trust-farmers-farmworkers-exploitation-agribusiness-low-pay-dangerous-working-conditions.

37. Dipak Subedi and Anil K. Giri, "Debt Use by U.S. Farm Businesses, 2012–2021," *Economic Information Bulletin*, no. 273 (June 2024), US Department of Agriculture, Economic Research Service, www.ers.usda.gov/publications/pub-details/?pubid=109411; Monika Ghimire, "Farm Sector Chapter 12 Bankruptcies in 2022 Lowest Since 2004," *Charts of Note*, US Department of Agriculture, Economic Research Service, May 3, 2023, https://ers.usda.gov/data-products/charts-of-note/chart-detail?chartId=106436.

38. US Department of Agriculture, Economic Research Service, "Farm Household Well-Being," updated February 6, 2025, https://www.ers.usda.gov/topics/farm-economy/farm-household-well-being.

39. Jingyi Tong and Wendong Zhang, "Iowa Farmland Ownership and Tenure Survey 1982–2022: A Forty-Year Perspective," Working Paper 23-WP 651, Center for Agricultural and Rural Development, Iowa State University, 2023, https://farmland.card.iastate.edu/files/inline-files/farmland-ownership-tenure-2022_0.pdf.

40. Anil K. Giri et al., "Commercial Farms Led in Government Payments in 2021," *Amber Waves*, May 15, 2023, www.ers.usda.gov/amber-waves/2023/may/commercial-farms-led-in-government-payments-in-2021; US Department of Agriculture, Economic Research Service, *Agriculture Resource Management Survey* (n.d.).

41. For Pollan on Naylor, see Michael Pollan, *The Omnivore's Dilemma: A Natural History of Four Meals* (Penguin, 2006), 32–48.

42. Maggie Koerth, "Big Farms Are Getting Bigger and Most Small Farms Aren't Really Farms at All," *FiveThirtyEight*, November 17, 2016, https://fivethirtyeight.com/features/big-farms-are-getting-bigger-and-most-small-farms-arent-really-farms-at-all.

43. See Henry Gordon-Smith, "Stories of the Year: What Does AeroFarms' Bankruptcy Signal for CEA's Future?," Food Institute, December 26, 2023, https://foodinstitute.com/focus/what-does-aerofarms-bankruptcy-signal-for-ceas-future.

Chapter 3: It's the Cow, Not the How

1. California Department of Food and Agriculture, *California Agricultural Statistics Review 2022–2023* (California Department of Food and Agriculture, 2024), www.cdfa.ca.gov/Statistics/PDFs/2022-2023_california_agricultural_statistics_review.pdf.

2. Peter Byrne, "Apocalypse Cow: The Future of Life at Point Reyes National Park," *Pacific Sun*, December 9, 2020, https://pacificsun.com/apocalypse-cow-the-future-of-life-at-point-reyes-national-park.

3. Hannah Ritchie and Max Roser, "Half of the World's Habitable Land Is Used for Agriculture," Our World in Data, February 16, 2024, https://ourworldindata.org/global-land-for-agriculture.

4. Monica Crippa et al., "Food Systems Are Responsible for a Third of Global Anthropogenic GHG Emissions," *Nature Food* 2 (2021): 198–209, https://doi.org/10.1038/s43016-021-00225-9.

5. "Per Capita Meat Consumption by Type, 2021," Our World in Data, accessed March 31, 2025, https://ourworldindata.org/grapher/per-capita-meat-type?time=2021.

6. Tony Weis, *The Ecological Hoofprint: The Global Burden of Industrial Livestock* (Zed, 2013).

7. Lindsey Sloat et al., "The World Is Growing More Crops—but Not for Food," World Resources Institute, December 20, 2022, www.wri.org/insights/crop-expansion-food-security-trends.

8. "Big Ag Is Draining the Colorado River Dry," Food & Water Watch, August 8, 2023, www.foodandwaterwatch.org/2023/08/08/big-ag-is-draining-the-colorado-river-dry.

9. Ellen K. Silbergeld, *Chickenizing Farms and Food: How Industrial Meat Production Endangers Workers, Animals, and Consumers* (Johns Hopkins University Press, 2016).

10. "New Study Finds Fast-Food Companies Spending More on Advertising, Disproportionately Targeting Black and Hispanic Youth," Fast Food FACTS, June 17, 2021, www.fastfoodmarketing.org/press.aspx.

11. Marco Springmann et al., "Analysis and Comparison of the Social Costs of Meat Consumption and Car Use," *Environmental Research Letters* 11, no. 10 (2016): 105002, https://doi.org/10.1088/1748-9326/11/10/105002.

12. Drew Pendergrass and Troy Vettese, *Half-Earth Socialism: A Plan to Save the Future from Extinction, Climate Change, and Pandemics* (Verso, 2024).

13. Franziska Funke et al., "Toward Optimal Meat Pricing: Is It Time to Tax Meat Consumption?," *Review of Environmental Economics and Policy* 16, no. 2 (2022): 219–239, https://doi.org/10.1086/721078.

14. Jeannette Cwienk, "German Supermarket Charges 'True Cost' of Foods," *Deutsche Welle*, August 2, 2023, www.dw.com/en/true-food-prices-germany-penny/a-66422126.

15. Ali Ladak and Jacy Reese Anthis, "Animals, Food, and Technology (AFT) Survey: 2020 Update," Sentience Institute, March 17, 2021, www.sentienceinstitute.org/aft-survey-2020.

16. Carey Sweet, "Family Behind Petaluma Duck Farm Pivots, Turns New Vision into Cookbook," *Sonoma Magazine*, November 2022, www.sonomamag.com/family-behind-petaluma-duck-farm-pivots-turns-new-vision-into-cookbook.

17. Marina Bolotnikova, "You're More Likely to Go to Prison for Exposing Animal Cruelty Than for Committing It," *Vox*, November 9, 2023, www.vox.com/future-perfect/23952627/wayne-hsiung-conviction-direct-action-everywhere-dxe-rescue-sonoma-county-chickens.

18. Jessica Fu, "Brown Gold: The Great American Manure Rush Begins," *Guardian*, February 2, 2023, www.theguardian.com/environment/2023/feb/02/manure-renewable-natural-gas-california.

19. Jan Dutkiewicz, "Why New York Is Suing the World's Biggest Meat Company," *Vox*, March 8, 2024, www.vox.com/future-perfect/2024/3/8/24093774/big-meat-jbs-lawsuit-greenwashing-climate-new-york.

20. Peter Scarborough et al., "Vegans, Vegetarians, Fish-Eaters, and Meat-Eaters in the UK Show Discrepant Environmental Impacts," *Nature Food* 4 (2023): 570, https://doi.org/10.1038/s43016-023-00795-w.

21. Walter Willett et al., "Food in the Anthropocene: The EAT–Lancet Commission on Healthy Diets from Sustainable Food Systems," *Lancet* 393, no. 10170 (2019): 447–492, https://doi.org/10.1016/S0140-6736(18)31788-4.

22. Jennifer Berg and Chris Jackson, "Nearly Nine in Ten Americans Consume Meat as Part of Their Diet," *Ipsos*, May 12, 2021, www.ipsos.com/en-us/news-polls/nearly-nine-ten-americans-consume-meat-part-their-diet; Kenny Torrella, "Americans Are Eating Less Meat. And More Meat. How?," *Vox*, November 21, 2024, www.vox.com/future-perfect/386374/grocery-store-meat-purchasing.

23. Jennifer Jacquet et al., "The Animal Agriculture Industry's Obstruction of Campaigns to Reduce Meat Consumption and Reform Factory Farming," *Climate Policy* 25, no. 4 (2025): 567–582, https://doi.org/10.1080/14693062.2025.2460603.

24. Vincenzina Caputo et al., "Do Plant-Based and Blend Meat Alternatives Taste Like Meat? A Combined Sensory and Choice Experiment Study," *Applied Economic Perspectives and Policy* 45, no. 1 (2023): 86–105, https://doi.org/10.1002/aepp.13247.

25. Emily Heil, "A Vegan Cheese Beat Dairy in a Big Competition. Then the Plot Curdled," *Washington Post*, April 30, 2024.

26. Chase Purdy, *Billion Dollar Burger: Inside Big Tech's Race for the Future of Food* (Portfolio, 2020).

27. P. D. Edelman et al., "*In Vitro*–Cultured Meat Production," *Tissue Engineering* 11, no. 5/6 (2005): 659–662.

28. Patrick D. Hopkins and Austin Dacey, "Vegetarian Meat: Could Technology Save Animals and Satisfy Meat Eaters?," *Journal of Agricultural and Environmental Ethics* 21, no. 6 (2008): 579–596, https://doi.org/10.1007/s10806-008-9110-0.

Chapter 4: Lunch Lady Politics

1. "Food Security in the U.S.—Key Statistics & Graphics," USDA Economic Research Service, updated January 8, 2025, www.ers.usda.gov/topics/food-nutrition-assistance/food-security-in-the-us/key-statistics-graphics.

2. Shawna Ohm, "Le Bernardin Chef Eric Ripert on How to Eat Like the 1%," Yahoo Finance, December 12, 2014, https://finance.yahoo.com/news/le-bernardin-chef-eric-ripert-on-how-to-eat-like-the-1-182133216.html.

3. "NYC True Cost of Living," United Way of New York City, 2023, https://unitedwaynyc.org/true-cost-of-living.

4. "USDA Food Plans: Monthly Cost of Food Reports," USDA Food and Nutrition Service, accessed March 31, 2025, www.fns.usda.gov/research/cnpp/usda-food-plans/cost-food-monthly-reports.

5. "Supplemental Nutrition Assistance Program (SNAP)—Key Statistics and Research," USDA Economic Research Service, accessed April 6, 2025, www.ers.usda.gov/topics/food-nutrition-assistance/supplemental-nutrition-assistance-program-snap/key-statistics-and-research.

6. Robert Paarlberg et al., "Keeping Soda in SNAP: Understanding the Other Iron Triangle," *Society* 55, no. 4 (August 2018): 308–317, https://doi.org/10.1007/s12115-018-0260-z.

7. Paarlberg et al., "Keeping Soda in SNAP."

8. "Kraft Warns on US Food Stamp Cut Plans," *Financial Times*, September 9, 2012.

9. "Consumer Expenditures in the New York Metropolitan Area—2022–23," US Bureau of Labor Statistics, Northeast Information Office, accessed April 6, 2025, www.bls.gov/regions/northeast/news-release/consumerexpenditures_newyork.htm.

10. Andrew D. Beck et al., "Association of a Primary Care-Based Mobile Food Pantry with Child BMIz," *Pediatric Obesity* 17, no. 5 (2022): e13023, https://doi.org/10.1111/ijpo.13023.

11. *2012 Harvest Report*, Farming Concrete, March 11, 2013, https://farmingconcrete.org/wp-content/uploads/2013/03/2012HarvestReport.pdf. See also Mara Gittleman, "Using Citizen Science to Quantify Community Garden Crop Yields," *Cities and the Environment (CATE)* 5, no. 1 (2012): article 4, https://digitalcommons.lmu.edu/cate/vol5/iss1/4.

12. Danielle Gallegos et al., "Food Insecurity and Child Development: A State-of-the-Art Review," *International Journal of Environmental Research and Public Health* 18, no. 17 (2021): 8990, https://doi.org/10.3390/ijerph18178990.

13. A. R. Ruis, "'The Penny Lunch Has Spread Faster Than the Measles': Children's Health and the Debate over School Lunches in New York City, 1908–1930," *History of Education Quarterly* 55, no. 2 (2015): 190–217, www.jstor.org/stable/24481675.

14. Quoted in Janet Poppendieck, *Free for All: Fixing School Food in America*, vol. 28 (University of California Press, 2011).

15. Sean Piccoli and Elizabeth A. Harris, "New York City Offers Free Lunch for All Public School Students," *New York Times*, September 6, 2017, www.nytimes.com/2017/09/06/nyregion/free-lunch-new-york-city-schools.html.

16. Kevin C. Mathias et al., "Food Sources and Diet Quality Among US Children and Adults: National Health and Nutrition Examination Survey, 2003–2018," *JAMA Network Open* 4, no. 4 (April 5, 2021): e215262, https://doi.org/10.1001/jamanetworkopen.2021.5262.

17. Piccoli and Harris, "New York City Offers Free Lunch."

18. Jessie Handbury and Sarah Moshary, *School Food Policy Affects Everyone: Retail Responses to the National School Lunch Program*, no. w29384 (National Bureau of Economic Research, 2021), www.nber.org/system/files/working_papers/w29384/w29384.pdf.

19. Andrew Knowlton, "Durham, N.C.: America's Foodiest Small Town," *Bon Appétit*, August 4, 2008, www.bonappetit.com/entertaining-style/holidays/article/america-s-foodiest-small-town.

20. "Housing Inventory: Average Listing Price in Durham County, NC," FRED, Federal Reserve Bank of St. Louis, accessed April 4, 2025, https://fred.stlouisfed.org/series/AVELISPRI37063.

21. "Food Programs," Durham Farmers' Market, accessed April 4, 2025, www.durhamfarmersmarket.com/our-programs.

22. Stacy Mitchell, "The Great Grocery Squeeze," *The Atlantic*, December 1, 2024, www.theatlantic.com/ideas/archive/2024/12/food-deserts-robinson-patman/680765.

23. Hunt Allcott et al., "Food Deserts and the Causes of Nutritional Inequality," *Quarterly Journal of Economics* 134, no. 4 (2019): 1793–1844.

24. "About," America's Healthy Food Financing Initiative, Reinvestment Fund, accessed April 4, 2025, www.investinginfood.com/about-hffi.

25. Matthew J. Salois, "Obesity and Diabetes, the Built Environment, and the 'Local' Food Economy in the United States, 2007," *Economics & Human Biology* 10, no. 1 (2012): 35–42, https://doi.org/10.1016/j.ehb.2011.04.001.

26. Bethany Schneider and Jordan Schott, *The Grocery Store Effect: How New Grocery Stores Impact Multifamily Rents in the Washington Metro Area* (Newmark Knight Frank, 2019).

27. Charles Courtemanche and Art Carden, "Supersizing Supercenters? The Impact of Walmart Supercenters on Body Mass Index and Obesity," *Journal of Urban Economics* 69, no. 2 (2011):165–181.

28. Charles Courtemanche et al., "Do Walmart Supercenters Improve Food Security?," *Applied Economic Perspectives and Policy* 41, no. 2 (2019): 177, https://doi.org/10.1093/aepp/ppy023.

29. See "Child Hunger & Poverty in Durham County, North Carolina," Feeding America, Map the Meal Gap, 2022, accessed April 4, 2025, https://map.feedingamerica.org/county/2022/child/north-carolina/county/durham; and "Durham County Food Security," Durham County Cooperative Extension, Durham County Government, accessed April 4, 2025, www.dconc.gov/county-departments/departments-a-e/cooperative-extension/food-security.

30. See Mary Helen Moore, "Durham Aims to Be Largest NC School District with Free Lunch," *News & Observer*, March 15, 2024, www.newsobserver.com/news/local/education/article286983120.html.

31. Akiya Dillon, "As Rents and Housing Prices Spike in Durham, So Does Unsheltered Homelessness," *INDY Week*, May 5, 2023, https://indyweek.com/news/ninth-street-journal/as -rents-and-housing-prices-spike-in-durham-so-does-unsheltered-homelessness.

32. "Durham Voters Approve $95M Affordable Housing Bond," City of Durham, November 6, 2019, www.durhamnc.gov/CivicAlerts.aspx?AID=2933.

Chapter 5: Workers and Eaters

1. Kathleen Kassel, "Agriculture and Its Related Industries Provide 10.4 Percent of U.S. Employment," US Department of Agriculture Economic Research Service, November 3, 2023.

2. Macy Stacher, "Waffle House Workers Challenge the Southern Economy," *American Prospect*, July 26, 2024, https://prospect.org/labor/2024-07-26-waffle-house-workers-challenge -southern-economy; Amy K. Glasmeier, "Living Wage Calculator," Massachusetts Institute of Technology, 2025, https://livingwage.mit.edu.

3. Joanna Fantozzi, "Waffle House Employees in South Carolina Go on Strike over Dangerous Working Conditions," *Nation's Restaurant News*, July 10, 2023, www.nrn.com/family-dining /waffle-house-employees-in-south-carolina-go-on-strike-over-dangerous-working-conditions.

4. Macy Stacher, "Waffle House Workers Challenge the Southern Economy," *The American Prospect*, July 26, 2024, https://prospect.org/labor/2024-07-26-waffle-house-workers-challenge -southern-economy.

5. "Child Labor in Agriculture," National Center for Farmworker Health, accessed April 6, 2025, www.ncfh.org/child-labor-fact-sheet.html. For state laws and FLSA provisions, see "Child Labor," US Department of Labor, accessed April 6, 2025, www.dol.gov/agencies/whd /child-labor.

6. "Despite Child Labor Laws, 8-Year-Olds Cut Asparagus," Opinion, *New York Times*, July 6, 1981, www.nytimes.com/1981/07/06/opinion/despite-child-labor-laws-8-year-olds-cut -asparagus.html.

7. "How Farm Workers' Rights Have Strengthened Since the 2008 Death of Pregnant 17-Year-Old María Isavel Vásquez Jiménez," KCRA, August 23, 2022, www.kcra.com /article/farm-workers-rights-pregnant-17-year-old-death-2008-maria-isavel-vasquezjimenez /40950637.

8. "Manure Pit Deaths Ruled Accidental," NBC4 Washington, May 25, 2012, www.nbc washington.com/news/local/manure-pit-deaths-ruled-accidental/1918142.

9. "Texas Fencing Contractor Fined $20K After 15-Year Old's Death," *Insurance Journal*, January 11, 2023, www.insurancejournal.com/news/southcentral/2023/01/11/702690.htm.

10. National Children's Center for Rural and Agricultural Health and Safety, "2022 Fact Sheet—Childhood Agricultural Injuries," Marshfield Clinic Health System, 2022, https://marsh fieldresearch.org/Media/Default/NFMC/National%20Childrens%20Center/2022_Child_Ag _Injury_Fact_Sheet.pdf.

11. "US Labor Department Proposes Updates to Child Labor Regulations," US Department of Labor, news release, August 31, 2011, www.dol.gov/newsroom/releases/whd/whd201 10831.

12. Gabriel N. Rosenberg, "A Painful Retreat on Child Labor," *Raleigh News-Observer*, May 2, 2012.

13. H.R. 4046, 118th Cong. (2023), www.congress.gov/bill/118th-congress/house-bill/4046 /text.

14. USDA Economic Research Service, "Ag and Food Sectors and the Economy," *Ag and Food Statistics: Charting the Essentials*, updated January 8, 2025, www.ers.usda.gov/data-products/ag-and-food-statistics-charting-the-essentials/ag-and-food-sectors-and-the-economy.

15. "Food Workers," Real Food Media, accessed April 6, 2025, https://realfoodmedia.org/issues/food-workers.

16. Tom Philpott, "Herman Cain's Enduring Lobbying Triumph," *Mother Jones*, July 30, 2020.

17. Matt Bruenig, "When McDonald's Came to Denmark," *Matt Bruenig Dot Com*, September 20, 2021, https://mattbruenig.com/2021/09/20/when-mcdonalds-came-to-denmark.

18. Jana Kasperkevic, "Were Fast-Food Workers Paid to Protest?," *Guardian*, September 8, 2014, www.theguardian.com/money/2014/sep/08/fast-food-workers-protest-strike-paid-mcdonalds.

19. Dominic Rushe, "'Hopefully It Makes History': Fight for $15 Closes in on Mighty Win for US Workers," *Guardian*, February 13, 2021, www.theguardian.com/us-news/2021/feb/13/fight-for-15-minimum-wage-workers-labor-rights.

20. Magick (@magick_the_hippie), "Join the 2 Million Dollar Club at Waffle House," TikTok video, November 27, 2022, 00:52 sec., www.tiktok.com/@magick_the_hippie/video/7170710104872946987.

21. Julian Roberts-Grmela, "US Starbucks Workers' Strike Expands to 11 States as Christmas Approaches," *Guardian*, December 23, 2024.

22. Matt Bruenig, "In a Union Triumph, the Seeds of Future Failure," Guest Essay, *New York Times*, August 21, 2024, www.nytimes.com/2024/08/21/opinion/starbucks-union-government.html.

23. Megan K. Stack, "Inside Starbucks' Dirty War Against Organized Labor," Guest Essay, Opinion, *New York Times*, July 21, 2023, www.nytimes.com/2023/07/21/opinion/starbucks-union-strikes-labor-movement.html.

24. "Union-Backing Shareholder Asks Starbucks to Disclose 'Anti-Union Spending,' Approaches SEC," Reuters, February 16, 2024, www.reuters.com/world/us/anti-union-efforts-cost-starbucks-least-240-mln-labor-group-tells-sec-2024-02-16.

25. "NLRB Judge Orders Starbucks to Give Union Employees Wage Increases in Catch-22 Case," Alerts and Updates, Duane Morris, October 31, 2023, www.duanemorris.com/alerts/nlrb_judge_orders_starbucks_give_union_employees_wage_increases_catch_22_case_1023.html.

26. "Attorney General James Secures $17.5 Million from DoorDash for Cheating Delivery Workers," New York State Office of the Attorney General, March 27, 2025, https://ag.ny.gov/press-release/2025/attorney-general-james-secures-1675-million-doordash-cheating-delivery-workers.

27. Laura Padin, "Prop 22 Was a Failure for California's App-Based Workers. Now, It's Also Unconstitutional," National Employment Law Project, September 16, 2021, www.nelp.org/prop-22-unconstitutional.

28. See Matthew Garcia, *From the Jaws of Victory: The Triumph and Tragedy of Cesar Chavez and the Farm Worker Movement* (University of California Press, 2012).

29. Kurtis Lee, "Farmworkers in California's Vineyards Fear the Loss of Their Jobs," *New York Times*, March 11, 2023.

30. *Farmworker Health in California: Health in a Time of Contagion, Drought, and Climate Change* (UC Merced Community and Labor Center, 2022), https://clc.ucmerced.edu/sites/clc.ucmerced.edu/files/page/documents/fwhs_report_2.2.2383.pdf.

31. Shay Myers, "Farmers Need Immigration Reform," *Washington Post*, October 21, 2021, www.washingtonpost.com/opinions/2021/10/21/myers-farmers-need-immigration-reform.

32. Steven Greenhouse, "In Florida Tomato Fields, a Penny Buys Progress," *New York Times*, April 24, 2014.

33. "The Fair Food Program—Consumer Powered, Worker Certified," Fair Food Program, accessed April 6, 2025, https://fairfoodprogram.org.

34. Mark Bittman, *Animal, Vegetable, Junk: A History of Food, from Sustainable to Suicidal* (Harvest, 2021), 270.

35. Diane M. Gubernot et al., "Characterizing Occupational Heat-Related Mortality in the United States, 2000–2010: An Analysis Using the Census of Fatal Occupational Injuries Database," *American Journal of Industrial Medicine* 58, no. 2 (2015): 203–211; Jill Rosenthal et al., "Extreme Heat Is More Dangerous for Workers Every Year," Center for American Progress, June 13, 2024, www.americanprogress.org/article/extreme-heat-is-more-dangerous-for-workers-every-year.

Chapter 6: In Praise of Processed Food

1. Priya Krishna, "A Brief and Buttery History of Libby's Pumpkin Pie Recipe," *Bon Appétit*, November 20, 2019, www.bonappetit.com/story/libbys-pumpkin-pie-recipe-history; Kimberly Holland, "Canned Pumpkin Actually Is Pumpkin, but Not the Kind You Carve," *Allrecipes*, October 8, 2021, www.allrecipes.com/article/whats-in-canned-pumpkin. The Dickinson squash is also sometimes called a Dickinson pumpkin, but, regardless, it's not the orange globe conventionally called a pumpkin.

2. Samuel D. Emmerich et al., "Obesity and Severe Obesity Prevalence in Adults: United States, August 2021–August 2023," *NCHS Data Brief*, no. 508 (September 2024), www.cdc.gov/nchs/products/databriefs/db508.html; "Diabetes Data and Research," Centers for Disease Control and Prevention, revised May 15, 2024, www.cdc.gov/diabetes/php/data-research/index.html; John Cawley et al., "Direct Medical Costs of Obesity in the United States and the Most Populous States," *Journal of Managed Care & Specialty Pharmacy* 27, no. 3 (March 2021): 354–366, https://pmc.ncbi.nlm.nih.gov/articles/PMC10394178.

3. Harvey Levenstein, *Paradox of Plenty: A Social History of Eating in Modern America* (University of California Press, 1993); Michael Pollan, *The Omnivore's Dilemma: A Natural History of Four Meals* (Penguin, 2006).

4. Ariana Brockington, "Gwyneth Paltrow Responds to Backlash over Her Wellness Routine," *Today*, March 17, 2023, www.today.com/health/gwyneth-paltrow-addresses-backlash-diet-rcna75545.

5. "The Joe Rogan Carnivore Diet," *Routines*, November 21, 2024, https://routines.club/routine/the-joe-rogan-carnivore-diet.

6. "United States Weight Loss Market Status & Forecast Report 2024," Finance.yahoo.com, May 30, 2023, https://finance.yahoo.com/news/united-states-weight-loss-market-084700.368.html.

7. William Kremer, "Lord Byron: The Celebrity Diet Icon," *BBC News*, January 2, 2012, www.bbc.com/news/magazine-16351761.

8. William Banting, *Letter on Corpulence, Addressed to the Public* (Harrison, 1863).

9. See Kyla Wazana Tompkins, *Racial Indigestion: Eating Bodies in the 19th Century* (New York University Press, 2012).

10. Julie Belluz, "How Dietary Supplements Evade Regulation—with Dangerous Results," *Vox*, April 8, 2015.

11. On the history of the FDA, see Daniel Carpenter, *Reputation and Power: Organizational Image and Pharmaceutical Regulation at the FDA* (Princeton University Press, 2014). On supplements, see Elizabeth Richardson et al., "What Should Dietary Supplement Oversight Look Like in the US?," *AMA Journal of Ethics* 24, no. 5 (May 2022): E402–409.

12. Erin Prater, "Tech CEO Defends Using His 17-Year-Old Son's Blood Plasma in Pursuit of Youth, Despite It Not Working," *Fortune*, July 13, 2023, https://fortune.com/well /2023/07/13/blueprint-ceo-bryan-johnson-defends-plasma-donation-son-youth-aging-longevity -brainstorm-tech-fortune-utah; Bryan Johnson, "Blueprint," accessed April 6, 2025, https://blue print.bryanjohnson.com.

13. Bryan Johnson (@bryan_johnson), "A lot of people ask me what I do about food when I travel. The first rule is this: food is guilty until proven innocent," X, December 2024, https://x.com/bryan_johnson/status/1863407743681368418.

14. Michael Pollan, "Unhappy Meals," *New York Times Magazine*, January 28, 2007, www .nytimes.com/2007/01/28/magazine/28nutritionism.t.html.

15. Mary Theoroda Weick, "A History of Rickets in the United States," *American Journal of Clinical Nutrition* 20, no. 11 (1967): 1234–1241.

16. Karen Clay et al., "The Rise and Fall of Pellagra in the American South," NBER Working Paper No. 23730 (National Bureau of Economic Research, revised May 2018), www.nber .org/system/files/working_papers/w23730/w23730.pdf.

17. "Life Expectancy in the USA, 1900–98," *Andrew Noymer's 1918 Flu Page*, accessed March 1, 2025, https://u.demog.berkeley.edu/~andrew/1918/figure2.html.

18. Russell W. Currier and John A. Widness, "A Brief History of Milk Hygiene and Its Impact on Infant Mortality from 1875 to 1925 and Implications for Today: A Review," *Journal of Food Protection* 81, no. 10 (2018): 1713–1722, https://doi.org/10.4315/0362-028X.JFP -18-164.

19. See Catherine McNeur, *Taming Manhattan: Environmental Battles in the Antebellum City* (Harvard University Press, 2014), 150–160.

20. Rachel Laudan, "A Plea for Culinary Modernism," *Jacobin*, May 22, 2015, https://jacobin .com/2015/05/slow-food-artisanal-natural-preservatives.

21. *Super Size Me*, directed by Morgan Spurlock (Hart Sharp Video, 2004), DVD; Billy Binion, "Super Size Me Was Not Groundbreaking Journalism," *Reason*, May 24, 2024, https:// reason.com/2024/05/24/super-size-me-was-not-groundbreaking-journalism.

22. Chris van Tulleken, *Ultra-Processed People: The Science Behind Food That Isn't Food* (Knopf Canada, 2023).

23. "Huel: Nutritionally Complete Food," Huel, accessed April 6, 2025, https://huel.com.

24. Paris Martineau, "Protein Powders and Shakes Contain High Levels of Lead," *Consumer Reports*, October 14, 2025, www.consumerreports.org/lead/protein-powders-and-shakes -contain-high-levels-of-lead-a4206364640.

25. "Healthy Eating Index Scores for Americans," USDA Food and Nutrition Service, accessed March 1, 2025, www.fns.usda.gov/cnpp/hei-scores-americans.

26. For a meta-analysis of eight studies, see Xueqin Gao et al., "Efficacy and Safety of Semaglutide on Weight Loss in Obese or Overweight Patients Without Diabetes: A Systematic Review and Meta-Analysis of Randomized Controlled Trials," *Frontiers in Pharmacology* 13 (September 13, 2022), https://doi.org/10.3389/fphar.2022.935823.

27. See Grace Niewijk, "Research Shows GLP-1 Receptor Agonist Drugs Are Effective but Come with Complex Concerns," UChicago Medicine, May 30, 2024, www.uchicagomedicine .org/forefront/research-and-discoveries-articles/research-on-glp-1-drugs.

28. Anne-Julie Tessier et al., "Optimal Dietary Patterns for Healthy Aging," *Nature Medicine*, 2025, https://doi.org/10.1038/s41591-025-03570-5.

29. Carlos A. Monteiro, "Nutrition and Health: The Issue Is Not Food, nor Nutrients, So Much as Processing," *Public Health Nutrition* 12, no. 5 (May 2009): 729–731, https://doi.org/10.1017/S1368980009005291.

30. Zhe Fang et al., "Association of Ultra-Processed Food Consumption with All-Cause and Cause-Specific Mortality: Population Based Cohort Study," *BMJ* 385 (May 8, 2024): e078476, https://doi.org/10.1136/bmj-2023-078476.

31. Fernanda Rauber et al., "Implications of Food Ultra-Processing on Cardiovascular Risk Considering Plant Origin Foods: An Analysis of the UK Biobank Cohort," *Lancet Regional Health—Europe* 43 (August 2024): 1–12, https://doi.org/10.1016/j.lanepe.2024.100715.

32. Mark Messina et al., "Perspective: Soy-Based Meat and Dairy Alternatives, Despite Classification as Ultra-Processed Foods, Deliver High-Quality Nutrition on Par with Unprocessed or Minimally Processed Animal-Based Counterparts," *Advances in Nutrition* 13, no. 3 (May 2022): 726–738, https://doi.org/10.1093/advances/nmac026. See also Virginia Messina et al., "Dietary Guidance on Plant-Based Meat Alternatives for Individuals Wanting to Increase Plant Protein Intake," *Frontiers in Nutrition* 12 (2025), https://doi.org/10.3389/fnut.2025.1641234.

33. Kevin D. Hall et al., "Ultra-Processed Diets Cause Excess Calorie Intake and Weight Gain: An Inpatient Randomized Controlled Trial of Ad Libitum Food Intake," *Cell Metabolism* 30, no. 1 (July 2019): 67–77, https://doi.org/10.1016/j.cmet.2019.05.008.

34. Ashley N. Gearhardt et al., "Social, Clinical, and Policy Implications of Ultra-Processed Food Addiction," *BMJ* 383 (October 9, 2023): e075354, https://doi.org/10.1136/bmj-2023-075354.

35. Rob Stein et al., "Layoffs Begin at HHS, Affecting Thousands of Staff and Leadership," NPR, April 1, 2025, www.npr.org/sections/shots-health-news/2025/04/01/g-s1-57485/hhs-fda-layoffs-doge-cdc-nih.

36. Marion Nestle, "Food Industry Funding of Nutrition Research: The Relevance of History for Current Debates," *JAMA Internal Medicine* 176, no. 11 (November 2016): 1685–1686, https://doi.org/10.1001/jamainternmed.2016.5400.

37. Madre Brava and Fern, *Making Ready-Made Meals Healthier and More Sustainable: The Role of EU Regulation*, March 2024, https://madrebrava.org/insight/madre-brava-fern-ready-meals-report.pdf.

38. See, for instance, Marco Springmann et al., "A Reform of Value-Added Taxes on Foods Can Have Health, Environmental and Economic Benefits in Europe," *Nature Food* 6 (January 2025): 161–169, https://doi.org/10.1038/s43016-024-01097-5.

39. Jonel Aleccia and the Associated Press, "Two Babies Infected with Rare Bacteria Sometimes Found in Powdered Infant Formula," December 8, 2023, *Fortune Well*, https://fortune.com/well/2023/12/08/abbott-baby-formula-recall-dead-child-brain-damage; Heather Vogell, "Unsanitary Practices Persist at Baby Formula Factory Whose Shutdown Led to Mass Shortages, Workers Say," *ProPublica*, April 4, 2025, www.propublica.org/article/baby-formula-abbot-sturgis-michigan-shortages-unsanitary-conditions-workers-say.

Conclusion: Feed the People!

1. Michael Grunwald, "Sorry, but This Is the Future of Food," *New York Times*, December 13, 2024, www.nytimes.com/2024/12/13/opinion/food-agriculture-factory-farms-climate-change.html.

Index

INDEX

Credit: Tim Atakora

JAN DUTKIEWICZ is an assistant professor at the Pratt Institute. He is a contributing writer at *Vox* and a contributing editor at *The New Republic*. He lives in Brooklyn, New York.

Credit: Harris Solomon

GABRIEL ROSENBERG is an associate professor at Duke University and a Senior Research Scholar at the Max Planck Institute for the History of Science in Berlin. He lives in Durham, North Carolina.

RAISING READERS
Books Build Bright Futures

Thank you for reading this book and for being a reader of books in general. We are so grateful to share being part of a community of readers with you, and we hope you will join us in passing our love of books on to the next generation of readers.

Did you know that reading for enjoyment is the single biggest predictor of a child's future happiness and success?

More than family circumstances, parents' educational background, or income, reading impacts a child's future academic performance, emotional well-being, communication skills, economic security, ambition, and happiness.

Studies show that kids reading for enjoyment in the US is in rapid decline:

- In 2012, 53% of 9-year-olds read almost every day. Just 10 years later, in 2022, the number had fallen to 39%.
- In 2012, 27% of 13-year-olds read for fun daily. By 2023, that number was just 14%.

TOGETHER, WE CAN COMMIT TO RAISING READERS AND CHANGE THIS TREND. HOW?

- Read to children in your life daily.
- Model reading as a fun activity.
- Reduce screen time.
- Start a family, school, or community book club.
- Visit bookstores and libraries regularly.
- Listen to audiobooks.
- Read the book before you see the movie.
- Encourage your child to read aloud to a pet or stuffed animal.
- Give books as gifts.
- Donate books to families and communities in need.

Books build bright futures, and **Raising Readers** is our shared responsibility.

For more information, visit JoinRaisingReaders.com

Sources: National Endowment for the Arts, National Assessment of Educational Progress, WorldBookDay.com, Nielsen BookData's 2023 "Understanding the Children's Book Consumer"